Genetic Theory of Reality

History and Theory of Psychology
Jaan Valsiner, Series Editor

Genetic Theory of Reality
Jones Mark Baldwin

The Oberservation of Human Systems
Joshua W. Clegg, editor

Genetic Theory of Reality

James Mark Baldwin

With a new introduction by

Jaan Valsiner

Routledge
Taylor & Francis Group
New York London

Originally published in 1915 by the Knickerbocker Press

Published 2010 by Transaction Publishers

Published 2017 by Routledge
711 Third Avenue, New York, NY 10017, USA
2 Park Square, Milton Park, Abingdon, Oxon OX14 4RN

First issued in hardback 2017

Routledge is an imprint of the Taylor & Francis Group, an informa business

Library of Congress Catalog Number: 2009037306

Library of Congress Cataloging-in-Publication Data

Baldwin, James Mark, 1861-1934.
Genetic theory of reality / James Mark Baldwin ; with a new introduction by Jaan Valsiner.
p. cm.
Originally published: New York : G. P. Putnam's Sons, 1915.
Includes index.
1. Reality. I. Title.

BD331.B2 2009
191–dc22

2009037306

ISBN 13: 978-1-4128-1085-2 (pbk)
ISBN 13: 978-1-1385-2423-1 (hbk)

To

ALL THOSE WHO FIND IN

ART

THE NOBLEST INSTRUMENT OF THE

SPIRITUAL LIFE

CONTENTS

PART II

DEVELOPMENT OF INTERPRETATION

CHAPTER IV

Early Racial Interpretation: Its Pre-logical Character.

CHAPTER V

Early Racial Interpretation: Its Positive Character

Contents

CHAPTER VI

The Religious Interpretation

CHAPTER VII

Religious Reality and Religious Negation

CHAPTER VIII

Logical Interpretation

Contents

CHAPTER IX

LOGICAL INTERPRETATION: MEDIATION THEORIES

CHAPTER X

IMMEDIATE THEORIES: I. THOSE BASED ON THE PRIMITIVE AND THE TRANSCENDENT

CHAPTER XI

IMMEDIATE THEORIES OF REALITY: II. THOSE BASED ON THE IMMEDIACY OF SYNTHESIS

Contents

CHAPTER XII

Results of the Historical Survey: the Demand for an Intrinsic Synthesis

PART III

ÆSTHETIC IMMEDIACY

CHAPTER XIII

The Intrinsic Synthesis, Æsthetic

CHAPTER XIV

The Æsthetic Interpretation

Contents

INTRODUCTION TO THE TRANSACTION EDITION

A PERSISTENT INNOVATOR: JAMES MARK BALDWIN REDISCOVERED

Jaan Valsiner

> *The psychology of the future will be social to the core; and its results, we surmise, will be revolutionary in logic, sociology, ethics, aesthetics, and religion—the disciplines which are built upon psychology.*
>
> *—James Mark Baldwin, 1905, p. 163*

JAMES Mark Baldwin was not a modest man. As one of the pioneers of psychology, sociology, and evolutionary thought in America his creativity seems without bounds. At his intellectual "reaching of age"—1880s and 1890s in the rapidly transforming society of the United States—which coincided with similar creativity of his peers—George Herbert Mead, John Dewey and others—their creative search followed the lead of Charles Sanders Peirce and William James. These were the times when new ideas were being pursued, old boundaries for thinking and believing transcended, as people moved into urban ways of living under the realities of increasing immigration and plurality of languages used in daily life. America was on the move into becoming relevant in the world—by creating new ideas rather than showing its political strength. The young generation of intellectuals was socially embedded in these change processes. Individually, the American thinkers raised above the social context—yet recognizing its primacy. The notion of the human mind as social first—and personal second—was widely shared at the time (Valsiner and van der Veer, 2000).

Baldwin left a legacy that is yet to be examined in its profound implications for science and the humanities. In some sense it paralleled that of his friend Charles Sanders Peirce whose semiotics becomes understood only a century later. Baldwin was trying to make sense of complex biological and social processes which only now come to the limelight of our biological sciences and—slowly but surely—re-emerge in psychology. Baldwin's focus on development—based on his observation of his own children and extrapolated to his general theoretical scheme—is fully in line with where our contemporary biological sciences are heading—as exemplified by the bounded flexibility of the work of the genetic system. The general principle of persistent exploration of the environment with the result of creating novelty—the core of Baldwin's theoretical system—is since 1960s the guiding idea in genetics. Likewise, our contemporary developmental science finds its roots in Baldwin's thinking (Cairns, Elder and Costello, 1996).

Baldwin's life, work—as well as personal fate—are a good example of the social processes of the American society of the latter quarter of the nineteenth century, internalized by an ambitious young man, and externalized in the form of the variety of activities with which he was involved, and on which he did not forget to reflect in his autobiographic expositions (Baldwin, 1926, 1930). He was an adamant organizer of new psychology—establishing laboratories in three universities in North America—yet he became uninterested in laboratory work very early in his career. He was constantly searching for answers to big questions in science—and in that search recognized no boundaries of the disciplines. He was an evolutionist, sociologist, psychologist, logician, and philosopher—consistently over his life. He was a pioneer in paternal child care—a proud father who publicly claimed how the pleasures of minding his daughters made the frustrations of parenting insignificant. And—last but not least—he defied the narrow-mindedness of the public morals—much to his own detriment.

Baldwin's contributions were major—and his ideas were well ahead of his times. Three of his major contributions need to be mentioned:

1) Demonstration of how human beings are in their nature "social beings"—which not only gave him international recognition but also a role in the history of sociology (Baldwin, 1897);
2) Establishing an alternative conceptualization of evolutionary theory (the "organic selection" model, or "Baldwin effect"—developed in parallel with Lloyd Morgan and Henry Osborn (Simpson, 1953; Weber and Depew, 2003); and,
3) Formulating a system of developmental ("genetic") logic. This system served as the basis for his immediate followers in developmental psychology—Jean Piaget and Lev Vygotsky among them. Baldwin could be considered to be the first to established multi-valued and deontic systems of logic (Valsiner, 2009).

It is not easy to summarize Baldwin's work. As one of the later commentators put it,

> ...Baldwin's earlier work was an epitome of the Darwinian epoch ... his later work was an expression of the general "philosophy of change," of the "taking time seriously," which followed upon this epoch (Urban, 1942, p. 305).

The present book—*Genetic Theory of Reality* (in English published in 1915, in French in 1918)—was the culmination of Baldwin's scientific career. It is an example of creating a system of aesthetic generalization. The beauty would keep human beings going—trying, and trying again—in the betterment of their society and themselves.

§ *I. Locating This Book: A General Theory of Developmental Aesthetics*

The present book—*Genetic Theory of Reality*—is the representative of the third contribution—the result of Baldwin's thinking in the first decade of the twentieth century. It was the culmination of his intellectual contribution to understanding development—be it of human beings, or of societies. It was worked out on the examples of aesthetics—as for many others of his time, the psychological relevance of art was a major intellectual challenge.

The present book followed three major volumes called *Thought and Things* (Baldwin, 1906, 1908a, 1911), each of which has clear links with the central idea in this final book on the role of affective synthesis in human lives. This was Baldwin's contribution to the social sciences—one that calls for a new look in our time. Being torn by the internal disputes within psychology between various *-isms* (behaviorism, positivism, cognitivism—to name a few) psychology in our days has largely overlooked complex affective phenomena. This area of ignoring has slowly developed over the twentieth century—as laboratory rats and pigeons pushing levers in cages could hardly be expected to be involved in philosophical or aesthetic contemplation. It is only now—the beginning of the twenty-first century—that the return to look at affect is again in vogue in the social sciences that actively claim to be interdisciplinary. Yet even today there is little effort towards basic philosophical generalization to art, religiosity, and human personal life philosophies. The popularity of the idea of "local knowledge" has fitted the empiricist credo of psychology where the post-modernist belief in impossibility of general knowledge has found its home. In contrast, Baldwin's *Genetic Theory of Reality* is dedicated to basic philosophical generalization—and therefore it is ahead of the psychology of emotions as it exists in our time.

§ 2. *Life as a Journey—Towards Personal Syntheses*

Born in South Carolina in 1861, by the 1880s Baldwin was a promising young man looking for a lifelong career. Like many American men of the 1880s, who later became psychologists, Baldwin at a time was not far away from becoming a Protestant minister. Going to get educated in a theological seminary was a way out of the religiously overcontrolled social world of late nineteenth-century America. It was an opportunity for young male minds ready for rebellion but still feeling the need to behave in socially proper ways (cf. the case of George Herbert Mead—Valsiner and van der Veer, 2000). Theological seminaries of the time accomplished the function of any society—to promote "the true" ways of knowing while simultaneously creating the basis for transforming those. What Serge Moscovici has later claimed about society at large was certainly true of the lives of young men in American theological seminaries in the 1880s:

> Our society is an institution which inhibits what it stimulates. It both tempers and excites aggressive, epistemic, and sexual tendencies, increases or reduces the chances of satisfying them according to class distinctions, and *invents prohibitions together with the means of transgressing them.* Its sole purpose, to date, is self-preservation, and it opposes change by means of laws and regulations. It functions on the basic assumption that it is unique, has nothing to learn, and cannot be improved. Hence its unambiguous dismissal of all that is foreign to it. Even its presumed artificiality, which might be considered a shortcoming, is taken, on the contrary, for a further sign of superiority, since it is an attribute of mankind (Moscovici, 1976, p. 149, added emphasis).

Social innovation in U.S. society has moved in waves. The critical minds in the American society of the turn of the twentieth century grew out from the theological education—meant to promote various kinds of prohibitions—by way of transgressing those promoted prohibitions. The Protestant theological seminaries in the United States—that were supposed to maintain the spiritual order of the society—by late nineteenth century were the breeding ground of new ideas and social innovations. This often entailed international openness—young Americans in the 1880s and later moved through European universities *en masse*—and even foreign languages of instruction out there seemed not to be a problem. Yet some decades before—in the mid-nineteenth century the ideas "foreign" were scrutinized carefully:

> The American philosopher ... must not attempt a *new* system of philosophy; but must seek to continue uninterruptedly, by improving it, the philosophy the race has always embraced, and as modified by the faith and practice of his own nation. In other words, the American philoso-

pher cannot transplant into his own country the philosophy of France or Germany, nor will it answer for him to seek to construct a philosophy for his countrymen from the French or German point of view. He must construct it from the English point of view, and continue English philosophy as modified … by Jonathan Edwards, our only American metaphysician, and by our peculiar civil, political, social and religious institutions. Our philosophy must be English philosophy Americanized, like the great mass of our population (Brownson, 1842, p. 358—cited via Rieber, 1998, pp. 208-209).

Metaphorically, this call for prudence was similar to building a barrier—a mental "Berlin Wall"—in the middle of the English Channel—at least for American thinkers. As its real example in the twentieth century, such wall could not be upheld permanently. Human beings cannot be kept in one place—they strive to leave their native villages, strive towards the unknown, explore new ideas, and reconstruct their life-worlds. It is the image of a ship—rather than of a church[1]—that leads ever-new generations of human beings to wander into the unknown.

The interests of human beings are usually drawn by not just the unknown, but of the curious realities. At the same time waves of fashions for the occult captivated the minds of not so philosophically minded Americans (Schmit, 2005)—mesmerism imported from France had an appeal. Promotion of a *status quo* guarantees that there will be—in some form and in some time—a break out of that very *status quo*. That general idea of development through action *against* the current order fits with Baldwin's notion of persistent imitation. In Baldwin's own retrospect upon the intellectual context of the United States,

The realism of the Scots—Dugald Stuart[2] and Reid[3]—served, in the universities and churches, to check and repel each succeeding tide of foreign speculation, up to the end of the nineteenth century, when a

> group of original thinkers arouse in America and gave the country for the first time a series of philosophical treatises of great originality and complete independence (Baldwin, 1926, p. 147).

He was certainly right—the efforts of William James, John Dewey, George Herbert Mead, G. Stanley Hall, and of his own were all testimony of breaking out from the theological closed circle, and create innovations beyond the expectations.

The Intellectual Grounds: Princeton (1881-1884) and James McCosh

Universities are a curious anomaly in any society—at our time, as well as in Baldwin's. They are secular seminaries—places for higher education where what "higher" means is constantly under dispute between the monastic self-organization of the university and the political structures of the society at large. What is being taught at a university is deeply ambivalent for the political powerholders of a society. On the one hand there is "right" and "useful" knowledge produced there—in the obscure world of academic robes and uneventful faculty meetings. On the other hand—new knowledge that can emerge there can be also "dangerous" to the survival of the political and ideological powers.

As Baldwin entered Princeton in 1881, he had no clear idea what to do in this seminary-now-university. It took a "social other"—the teacher—who guided his mind towards becoming involved in science. That "other" was nobody else but the first president of the university—James McCosh—a representative of Scottish intuitional realism. The Scottish influence in U.S. seminaries and colleges was overwhelming in the 1860s—it suggested that science refer to "any seemingly systematically organized body of knowledge, and was initially treated synonymously with theology and philosophy" (Spilka, 1986, p. 3). While in Europe psychology emerged from physiology and philosophy, in North America it was the theological breeding ground that catalyzed the development of the newly independent science.

Under McCosh, Princeton went through a transformation—psychology displaced orthodox theology in this little College of New Jersey (Maier, 2005). Within that social context, it was the personal influence of McCosh during Baldwin's undergraduate years that guided him into psychology. As Robert Wozniak has pointed out

> It is impossible to overestimate the extent to which Baldwin was influenced by McCosh, both as man and as thinker. McCosh was a man of balance; and much of Baldwin's work can be and should be read as reflecting this same striving to maintain an integrative equilibrium between opposing tendencies. McCosh was transitional, willing to entertain the new ideas of evolution and of an experimental, physiological psychology, and it was through McCosh that Baldwin received his first introduction to both these innovations. Finally, McCosh was, first and foremost, a theistic epistemologist, concerned to ground faith and morality on rational intuition, self-consciousness, and natural realism. From McCosh, Baldwin adopted his unswerving commitment to reason, his belief in the existence of separate and primary order of mental facts in consciousness, his realistic metaphysics, and his own concern with the psychological justification of morality (Wozniak, 1998, p. 433).

Baldwin himself was appreciative of McCosh's role in his education—moving him gradually away from theology towards philosophical and psychological issues. Yet as is usual for sons—even intellectual ones—he resisted the direct suggestions of his "father figure" not to get involved in the study of the philosophy of Spinoza for his Ph.D.

Setting the Sails: First Visit to Germany

When he graduated from Princeton, Baldwin had the opportunity to visit Germany on a post-graduation fellowship. Serious study of psychology in those days meant a stay in Germany, for at least some time. Baldwin went there for three semesters in 1884-85. First he spent a semester in Leipzig in Wilhelm Wundt's laboratory and listened to his lectures, and, in his words, "was caught in the movement" (Baldwin, 1930, p. 2) for "new psychology." Yet Baldwin did not follow Wundt. Instead, he moved to appreciate Spinoza (through his experiences in Berlin with Friedrich Paulsen[4]), and spent his third semester in Freiburg with Carl Stumpf). Baldwin's main intellectual benefit from the study trip was the introduction to Spinoza's philosophy in Berlin (see Baldwin, 1926, chapter 3; also Baldwin, 1902a, chapter 2). The work of Hermann Lotze[5] likewise became a general source of inspiration for Baldwin's ideas.

Back in the United States—Princeton (1885-1887), and Lake Forest (1887-1889)

Baldwin returned to the United States as an instructor at Princeton in 1886 to teach French and German, followed by an appointment in philosophy and psychology in 1887 at Lake Forest University in Illinois—an extension of Princeton to the Midwest. This coincided with his marriage—to the daughter of the president of Princeton Theological Seminary (Helen Green). He introduced the new German psychology into the United States by translating Theodule Ribot's book *Psychologie allemande contemporaine* into English (Ribot, 1886). By 1887, Baldwin's basic worldview was in place (Wozniak, 1998, p. 439). Likewise his theory of synthesis—in apperception, conception and volition—followed suit in the 1890s (Baldwin, 1930, p. 3).

Stay in Canada (1889-1893)

Baldwin needed a more solid university position—and got it in Toronto. Moving North was good—it turned out to be a crucial period in Baldwin's theoretical development. After a large local-Canadian controversy about a professorship offered to a "foreigner"—from the inferior lands of the United States,[6] Baldwin moved to Canada to take on the professorship in logic and metaphysics at University of Toronto (Green, 2004). The link with his Scottish philosophical background from Princeton proved to be the key for Baldwin's becoming accepted in Toronto—yet it had to be masked by his recent learning in Europe with Lotze (Green, 2004, pp. 146-147). In these few years in Toronto—1889 to 1893—Baldwin's intellectual affiliations developed further in the Francophile direction. It was in Toronto that he established the first psychology laboratory on the British soil, where he was exposed to psychic phenomena of esoteric kinds, where he did most of his child observations (on his own daughters, while babysitting them—Baldwin, 1892a), and where he grew uninterested in organizing an endless numbers of empirical studies.

The focus on the paranormal psychological phenomena was a challenge to Baldwin. As a philosopher he could take interest in such phenomena from a comfortable safety position of one's armchair. As a psychologist one is thrown into the ring of social demands of explaining the supernatural. These phenomena—"animal magnetism" and cases of hypnotic and post-hypnotic suggestion—were widely discussed among late nineteenth-century intellectuals (Valsiner and van der Veer, 2000, chapters 2 and 3). Baldwin was less than enthused by the myriad of "cases of visions of phantoms or ghosts, premonitions of death" (Baldwin, 1926, p. 45) which he was asked to examine. Baldwin had to diplomatically deal with these "real-life phenomena" that were brought to him for scientific scrutiny.

The scientific scrutiny of phenomena close to those of mesmerism was happening in France. Baldwin visited Paris (the clinic of Charcot at Salpetriere) and Nancy (the center of hypnotism of Hypolite Bernheim) in 1892 (his accounts of the trip are recorded in Baldwin, 1892b, 1892c). Since then, Baldwin's closest intellectual partner was Pierre Janet. His thinking benefited from the work of Alfred Binet, Gabriel Tarde, Theodore Flournoy (to whom he dedicated the first volume of his *Thought and Things*—Baldwin, 1906), Edouard Claparède and Henri Bergson (Baldwin, 1926, 1930). This visit set the stage for his later life in exile in England and France—away from the powerhouses of psychology in post-World-War I Europe, but very deeply connected with the best of French intellectual circles. Regular lunch meetings with Pierre Janet in Paris were obviously more enjoyable than presiding some professional society of psychologists could ever be.

The second major feature of the Toronto-period in Baldwin's life that is important to mention as of relevance for his ideas, is the benefit Baldwin gained from direct contact with psychological phenomena in the practical context of taking care of his two daughters (Helen, born 1889; Elizabeth, born 1891). Baldwin could be considered as one of the forebears of our contemporary fashion of emphasizing the role of fathers in the child care process. For science, his child care practices provided the basis for theoretical elaboration of his key concepts—*circular reaction* and *persistent imitation*—which remain the core of his theory of aesthetic synthesis in the present book.

The key concept for understanding development—persistent imitation. Persistent imitation is central for understanding Baldwin's efforts to make psychology both developmental and social. It is a form of "imitation" that goes beyond the model it "imitates"—by introducing new features to the "copy" of the "model." Thus to call it *imitation* is in some sense a misnomer—as it defies the common sense assumption of any "imitation" being that of the model—rather than beyond the model. *Persistent imitation is an act of re-construction of a model given—in a new form*—which is then used in action. Such use can create a recursive feed-forward to produce novelty—the first creative imitation becomes a model for the next, etc—so the novelty production can escalate.

The reasons for Baldwin's need to create this concept can be found in his efforts to explain the future-oriented developmental processes at the level of the individual:

> Suppose at first an organism giving random reactions, some of which are useful; now for development the useful reactions must be repeated, and thus made to outweigh the reactions which are damaging or useless. Evidently if there are any among the useful reactions which result in immediate duplication of their own stimulus, these must persist, and on them must rest the development of the organism. These are

> the imitative reactions. Thus it is that *a thing in nature once endowed with the reacting property might so select its stimulations as to make its relations to its environment means to its own progress*: imitative reactions, as now defined, being the only means to such selection (Baldwin, 1894, p. 29; emphases added).

Here the need for inventing some notion of imitation becomes obvious as a deductively created object of study. The functionalist and instrumentalist stance of Baldwin's is clearly visible (e.g., reference to "usefulness" in general). The developmental orientation helped him to extend the sequence of a stimulus and reaction into a time-extended "circular reaction"—in which the organism's reaction to a stimulus becomes the next stimulus for the organism, leading to the next reaction, and so on. The organism begins to construct non-random experiences on the basis of such circular reactions. The reactions become instruments, and thus begin to participate in the developmental process. The imitative reactions are thus the key to understanding the process of development (Baldwin, 1895). It parallels John Dewey's move to turn the "reflex arc" into a "reflex circle" (Dewey, 1896), as well as Jakob von Uexküll's "functional circle" in his *Umweltlehre* (Chang, 2009; Kull, 2001; von Uexküll, 1926). Baldwin's work on human development was the basis for Jean Piaget and Lev Vygotsky years later.

Back in Princeton—The Third Stay (1893-1903)

Baldwin's third Princeton stay was the time for his charting out the social origins of the self—culminating in his celebrated *Social and Ethical Interpretations in Mental Development* (Baldwin, 1897). That book was his main contribution to sociology—which in the 1890s was rapidly developing in North America. It is therefore not surprising that Baldwin was later considered—by American sociologists—to be one of the founders of that discipline in North America.

In his sociological psychology of development, Baldwin transcended the limits of static ontology of being—"I am X" by assuming a stance of becoming ("I am X while becoming Y") which works through moving through the social encounter ("I am X while becoming Y through the encounter with Z of you"). The psychological uniqueness of each person is the proof of the social origins of the self that develops through the relating with others:

> *[A] man is a social outcome rather than a social unit.* He is always in his greatest part, also some one else. Social acts of his—that is, acts which may not prove anti-social—are his *because they are society's first;* otherwise he would not have learned them nor have had any tendency to do them (Baldwin, 1897, p. 87).

Baldwin's consistently developmental perspective made it possible to reconcile the uniqueness of person—who imitates socially given models constructively—and the knowledge base of the *socii.* He evoked the image of a boomerang in the relations of the person and the society ("When he acts quite privately, it is always with a boomerang in his hand; and every use he makes of his weapon leaves its indelible impression both upon the other and upon him"—ibid, p. 88).

The notion of self-as-boomerang-thrower fits with the basic developmental ethos that Baldwin carries systematically through in all of his thinking. The self and the others continuously relate with one another, feeding forward into one another's individual development. According to Baldwin, such dialectic relation could be described like this:

> [M]y thought of self is in the main, as to its character as a personal self, *filled up with my thought of others, distributed variously as individuals; and my thought of others, as persons, is mainly filled up with myself.* In other words, ... the self and the alter are to our thought one and the same thing (Baldwin, 1897, p. 343, added emphasis).

Being "one and the same thing" is not an ontological statement of being—but a dynamic statement of directed movement. In order to "be myself" I "fill my thinking" with what "others" might—or do—think of me, and in return I communicate to them what I happen to think of them. The "boomerang of thought" is thrown towards the "alter" in order to return to myself—and the myriads of similar communicative "boomerangs" are thrown from all others towards me—only to return to their "throwers."

Personally internalized social selectivity. From the heterogeneity of the person's social environment follows the need for selective treatment of that heterogeneity by the person himself. It is only the person who remains continuous within his- or herself, across the myriad of constantly changing environmental conditions. A previously established "schema" (see Baldwin, 1908b, p. 184) allows the person to become selective as to the variety of presently actual environmental inputs. According to Baldwin,

> the individual *gradually builds up internally* the criteria of selection; and as his experience extends even more widely afield from the brute resistances, strains, and contacts with things, he becomes a more and more competent judge for himself of the value of variations in his thoughts (Baldwin, 1898, p. 17).

However, the person's internalization of social experiences entails development of autonomy and such autonomy is itself socially constituted in any form that it takes. Thus, the developing person

> comes more and more to reflect the social judgment in his own systematic determination of knowledge; and there arises within himself a criterion of private sort which is in essential harmony with the social demand, because genetically considered it reflects it. The individual becomes a law unto himself, exercises his private judgment, fights his own battles for truth, shows the virtue of independence and the vice of obstinacy. But he has learned to do it by the selective control of his social environment, *and in this his judgment he has just a sense of this social outcome* (Baldwin, 1898, pp. 19-20).

It is obvious that the social nature of a person is expressed in his personal individuality, rather than in the mere direct mirroring of the social world surrounding the person. The latter is already rendered impossible by its high heterogeneity (which triggers the need for "systematic determination" of the new knowledge by way of internalized selection mechanisms—cognitive schemata; see Baldwin, 1898, p. 10). Baldwin saw the person <—> society relationships as a process of particularization of general meanings by persons in specific contexts on the one hand, and generalization (of persons' thought-variations by society) on the other (Baldwin, 1897, part V). Baldwin's work at the time provided an impetus for sociology in America (Giddings, 1899).

The Organic Selection Theory

In the 1890s serious thinkers were truly interdisciplinary. Aside from his contributions to sociology—based on his psychological observations—Baldwin was also contributing to evolutionary theory. Baldwin's ideas grew in parallel to those of his contemporaries—C. Lloyd Morgan and Henry Osborn. Their co-created organic selection theory—introduced in 1896—constituted an effort towards overcoming this limit of the selection notion in evolutionary theory. It built on solid intellectual roots of their predecessors (Bateson, this book).

Morgan's, Osborn's and Baldwin's goal was to find a third way between the ideology of full genetic determinism and its equivalent on the side of environmental causation—unlimited Lamarckianism. Looking back at this goal from the viewpoint of a hundred years of genetics since the re-discovery of Mendel's laws such goal could only stand out in the most positive fashion. As the world commemorates two dates in 2009—150 years from Darwin's *Origin of the Species* and 200 years of Lamarck's *Philosophie Zoologique* we can see how contemporary protein genetics brings to daylight some of the processes Lamarck could only intuitively grasp. The "third way" of the three organic selection theorists of 1896 is moving into the center of our science.

The development of organisms is given by the basic genetic blueprint (the DNA→ proteins transmission)—yet its specifics are modifiable as novel features of the environment are encountered (RNA→ DNA reverse transcriptions). In recognizing this path on biological adaptation—well before the actual genetic mechanisms, as well as the very notion of Mendelian discoveries—became known, Baldwin's creative mind could predict where the science of biology would be a century later. Flexibility of the developing organism—based on the perpetual uncertainty of the immediate future—is the basis for survival. The psychological world is similar—children develop under the guidance of their parents—yet in ways that transcend the ways parents tried to guide them into. Baldwin's observations of children's—his own daughters'—development provided him with theoretical insights that he transferred to his evolutionary theory.

Development—be it of viruses, humans, or societies—is an act of exploration that results in new, previously unencountered, forms of being. A simple idea—yet it has taken biological and social sciences around two centuries to accept it.[7] Interestingly, Baldwin's version of it was based on his observations of the play of his own young children from where it was extrapolated to the abstract notion of "organic evolution" (Baldwin, 1902b). Every parent of course observes one's children—but there are very few who derive a basic principle from those observations.

The crucial notion in this "third way" was the notion of *convergence of direction* between the acquired innovative behavioral patterns and those guaranteed by genetic programs. As Baldwin had summarized earlier,

> In any given series of generations, the individuals of which survive through their susceptibility to modification, there will be a gradual and cumulative development of coincident variations under the action of natural selection. The adaptive modification acts, in short, as a screen to perpetuate and develop congenital variations and correlated groups of them. *Time is thus given to the species to develop by coincident variation characters indistinguishable from those which were due to acquired modification, and the evolution of the race will proceed in the lines marked out by private and individual adaptations.* It will appear as if the modifications were directly inherited, whereas in reality they have acted as the fostering nurses of congenital variations (Baldwin, 1897, p. 558, added emphasis).

Since any evolutionary discourse pertains strictly to the level of phylogeny, the role of "fostering nurses of congenital variation" applies to inter-generational accumulation (or maintenance) of the heterogeneous set of new forms. That maintenance requires non-genetic transfer of inventions from one generation to the next—through principles of re-construction. If such mechanisms are worked out, the species gains valuable phylogenetic time for selective "genetic assimilation" of the acquired characteristics (Simpson, 1953). Lamarck and Darwin were indeed both fitting into this scheme.

The Years at Johns Hopkins (1903-1909)—Creating Genetic Logic

After disagreements with Woodrow Wilson—then the president of Princeton University—Baldwin moved to Baltimore in 1903 to lead the re-establishment of a psychology laboratory there (his third to establish at different North American universities). By that time he had lost most of his earlier interest in laboratory research, and was moving increasingly towards the development of his general system of developmental thought—his genetic logic (Baldwin, 1906,[8] 1908a, 1911), the culmination of which is the present book. It would be fair to add that psychology in the United States around that time moved away from Baldwin's intellectual world. The three volumes of *Thought and Things*, and their sequel—*Genetic Theory of Reality*

> caused few reverberations in the American intellectual community, since the *Genetic Logic* reeked of the argot of deep idealistic philosophy, particularly epistemologies of Lotze, Wundt, and Bradley, while the philosophic temper of America had warmed up to pragmatism and empiricism. And for his fellow psychologists, who were trying to cut their ties to philosophy and to begin speaking the language of mathematics, Baldwin's metaphysical schemes and uncontrolled style proved quite incomprehensible (Richards, 1987, p. 496).

A similar story happened around the transposition of George Herbert Mead—within the very same University of Chicago over three decades—from psychology to philosophy (see Valsiner and van der Veer, 2000). American psychology turned decidedly anti-philosophical—claiming to look for useful knowledge, rather than be involved in contemplative perspective taking towards the world. Pragmatism won—creating a powerful factory of empirical research that became detached from addressing basic human question.

Basis in logic. Baldwin set his inquiry up within the root metaphor of logic. He had good reasons for doing that—logic was the "gold standard" of scholarly thinking around 1900s, and one of his closest fellows—Charles Sanders Peirce—worked within a similar frame.[9] At the same time he was ambiguous as to the difficulty of fitting the notion of logic—its deductive and inductive versions—to psychological phenomena. From time to time his coverage turned to practice of thinking that was beyond either of these versions—moving in parallel with Peirce's efforts to develop the method of abduction as a third way. Baldwin's focus was largely out of touch with the development of logic in philosophy and mathematics in the twentieth century (Valsiner, 2009). After his death, Wilbur Urban commented:

> A genetic theory of logic was a natural development from a genetic theory of mind. The revaluation of intellect and logic, in the light of Darwinian principles—of instrumental logic as developed by John Dewey and of Intuitionism as developed by Bergson—were prime concerns of Baldwin's more important contemporaries in philosophy. In this revaluation Baldwin played a conspicuous part. The extreme positions of his contemporaries tended to overshadow his more moderate and mediating views. The development of logic in the purely formal, mathematical direction has temporarily withdrawn interest from the philosophical problems of logic as contemplated by Baldwin. But these problems must inevitably come to the foreground again, and it is, perhaps, not too much to say that his careful analysis and his magnificent picture of logic and reason in its total functioning, will again have something to say to philosophy (Urban, 1942, p. 305).

Urban's prophecy proves itself right—in the resurgence of developmental science in the 1990s (Cairns, Elder and Costello, 1996), and in breaking through the mist of statistical treatments of phenomena into new versions of qualitative mathematics, including new formal logics (Valsiner, 2009; Valsiner and Rudolph, 2008).

Baldwin's goal in creating his genetic logic was to "interpret the processes usually handed over to formal logic in terms of the movements of actual knowledge" (Baldwin, 1908, p. x). There was clear closeness of Baldwin's effort to that of the "Graz group" of psychology—led by Alexius Meinong (Albertazzi, Jacquette and Poli, 2001; Grossmann, 1974), and to other key figures in Brentano direction (Anton Marty). Baldwin (1906, p. viii) recognized that explicitly—and made an apology of his own work completed by the time he could get acquainted with it. This link was further evident—yet without explication—in the second volume of *Thought and Things* (Baldwin, 1908, parts 2 and 3)—and again briefly acknowledged in volume 3 (Baldwin, 1911a, p. 275). Obviously, Baldwin saw the parallel efforts of the "Graz group" to handle problems similarly to his—yet distanced his own coverage from the Austrian tradition. Priority claims were not foreign to Baldwin's ego.

Beginnings of developmental science. Baldwin was the originator of the direction of thought that by our time has acquired its own autonomy under the label of *developmental science* (Cairns, Elder, and Costello, 1996). He understood the great divide in the meaning systems of developmental and non-developmental kinds. He understood then what most psychologists today still fail to do—it is useless to import the non-developmental notions of causality into the new "genetic [=developmental] science":

> We must be free from all constructions drawn from the strictly a-genetic sciences in which the causal sequence is the typical one. The birth of a new mode in the psychic life is a *"progression" from an earlier set of conditions, not the effect of these conditions viewed as cause*; and this is equally true of any new genetic mode, just so far as the series in which it appears is really genetic at all (Baldwin, 1906, p. 29, emphasis added).

Baldwin was caught in-between the glorifiers of the expressivity of natural language for psychology (e.g., William James) and those who would prefer the precision of mathematics (Charles S. Peirce, for one). While recognizing the confusions coming from the common language, Baldwin was not ready to give in to full formalization either (cf. Baldwin, 1930, p. 26). Thus, in his presentation

> logic ... may be applied to the processes of mind in general, all recognized as being parts of one's continuous movement. This had analogies in expressions already in use, such as "logic of experience," logic of history" etc. *Genetic logic was, in my usage, the term adopted to designate the body inside or psychic process in which mental development takes place. Within this logic, all the varied special motives of adaptation, opposition, assimilation, etc. uncovered in the detailed researches, show themselves in the phenomena of personal and social progress* (Baldwin, 1930, pp. 11-12, added emphasis).

Baldwin was eager to solve the problem of relations between human affective and rational sides. He was eager to look at processes of affective generalization and idealization (as parts of the affective processes) in relation to their rational-logical counterparts:

> Affective logic is a process proceeding by the mediation of ends through means; *its result is always in the domain of an interest or value.* On the

> other hand, that of cognition is in the domain of truth. But there are all sorts of criss-crossings and interferences between the two, the process of truth-seeking rarely being free of influence from the tendencies of feeling and interest which assert themselves when most unexpected. *Here the "will to believe" shows itself actively, by the intrusion of interest; it finds value at the end of a process which claims to issue solely in the establishment of truth* (Baldwin, 1930, p. 18, added emphases).

Baldwin's dialogue here is with William James' pragmatism—taken to the extreme in the manifesto of "the will to believe."[10] Baldwin also antedates the invention of "cognitive heuristics" in the 1970s—evident in the work of Amos Tversky, Daniel Kahnemann, and Gerd Gigerenzer. The recognition that the affective side of human beings intrudes in the cognition to undermine its movement to "truth" and produce another value is the "bias" the heuristics produce.

Affective processes are teleological. They express desired values and beliefs in ways that relate the presently established belief with one desired for the future, or that are simply considered true. This teleology can occur through processes of affective generalization, ejection, and idealization Affective generalization entails the distribution of a particularly established emotional tone from the object (or context) of its original emergence, to other objects (and contexts). This process is complemented by ejection:

> "the self ... embodied in ... mass of general interests, ejects itself semblantly [empathically] or imaginatively into other selves, and so establishes meanings of common interest, conformity and practice" (Baldwin, 1911, p. 94).

Thus—a person is social not by way of "succumbing" to "social influences" from outside, but by actively feeling into the "social others" on the basis of one's own affective understanding of the world. It involves idealization—which is a by-product of constructive internalization. In the process of establishing personal meanings of objects, the object becomes "charged with further unfulfilled meaning" (Baldwin, 1906, p. 233).

Persons always create ideal meanings of objects, aside from—and on the basis of—their commonly shared meanings. These ideal meanings orient the person in relation to the given object, as well as to other objects:

> the ideal meaning resides (1) in the objects constructed out of certain materials—these materials ... *not allowing any construction which does not have something of the ideal meaning*—and (2) in the treatment of other materials in the same way, that is, *as if they also had the characters which normally take on this meaning* (Baldwin, 1906, p. 235).

The person establishes an idealized view of one object in his or her encounters with it, and then transfers it to further encounters with other objects. In its function, the idealized meaning limits the set of possible affective relations to these objects. It serves as the mediating device that imbalances any equally-valenced disjunctive situation ("X is equally either Y or Z"), turning it into a case of affective dismissal of one of the options ("X must be Y, even if it can be both Y or Z").

Centrality of empathy (sembling). It is in the realm of discussing play and art that Baldwin connected his theorizing with the issue of *Einfühlung* of Theodor Lipps. Baldwin's translation of that term into English is a good example of his liking for terminological inventions: he translated it as *sembling* ("to semble" = to make like by imitation; Baldwin, 1906, p. 122). Of course, in the decades after Baldwin the competing translation (suggested by E. B. Titchener and J. Ward; Baldwin, 1911, p. 167) of that concept—*empathy*—has become accepted in psychology, albeit with all the confusions that *Einfühlung* brought with it in the beginning of the twentieth century.

Originally focusing on the "feeling-into" an object, Baldwin's version of sembling translates the *Einfühlung* into his imitation-centered conceptual system:

> Broadly understood, the process of Sembling consists in the reading-into the object of a sort of psychic life of its own, in such a way that the movement, act, or character by which it is interpreted is thought of as springing from its own inner life (Baldwin, 1906, p. 124).

Sembling thus entails projection upon one another of the oppositions that persist within one's own "inner" <—> "outer" relationships, as those are construed within the mind's "inner imitation." It was claimed to be present in both play and art. The "self sembles itself"—the child takes on new roles, attaches its personal understanding to their external demands, and that leads to further development of the self.

Abandoning the United States: Mexico and Europe

After having been pushed out from the U.S. academic world for his "moral misdemeanor" in 1908 (see Richards, 1987 for the description), Baldwin moved between Mexico City and Europe, with a focus on teaching at the Universidad Autonoma de Mexico. His visits to Mexico were short (Gallegos, 1980) and were not able to contribute to the development of psychology in Mexico—as the revolution interfered (Hurley, 2001).

Most of the writing of the third volume of *Thought and Things*—subtitled *Interest and Art Being Real Logic. I. Genetic Epistemology* (Baldwin, 1911) happened at this time of Baldwin's entrance into the uncertain life space as a rejected outsider. Dedicated to the memory of William James[11] (who had died in August 1910 and whose seat as corresponding member of the French Academy of Moral and Political Sciences Baldwin took over in 1910), the third volume of the series deals with the imaginative side of human minds. The focus on art—developed further in the present book by 1915—is already present in 1911. That focus gets particular coverage in the latter part of the 1911 book—chapters 10-14 that were originally published in Spanish in Mexico City (Baldwin, 1911, p. 156).

The emergence of affective value. Baldwin in 1911 addressed the emergence of higher affective processes—interests and value. In the search for ideal—affectively affirmed—human beings use

> the role of the imagination to seize upon an idea, thought, or another content, "semblantly," in the pursuit of a special interest; to treat it as being *more than it is.* This is seen in the experimentation or schematism of the development of knowledge, in the "make-believe" of play, in the self-projection or *Einfühlung* of the aesthetic (Baldwin, 1911, p. 129).

The latter famous claim of thinking involving "going beyond the information given" is preceded by Baldwin's look at the feed-forward—or in his terms—*feel*-forward—processes.

Feel-forward mechanisms and higher mental functions. Sembling as was shown above is based on imagination, which is oriented towards possible future events. It is viewed by Baldwin in his characteristic prospective way, emphasizing directedness towards ideal goals:

> the sort of meaning known as ideal, due to an imaginative *feeling-forward,* has an essential place in the development of the affective life. The entire movement of cognition and feeling alike has not only the

> interest and intent to conserve its data and preserve its habits, but also the interest and intent to achieve, to learn, to adapt, to acquire, to *feel-forward* (Baldwin, 1911, p. 125).

Sembling is the obligatory mechanism for the feeling-forward process. It is here that the emergence of higher psychological functions—meaningful personal willfulness—is under the challenges of the task demands. Such higher psychological functions entail the creation and extension—in time and space—of personal life philosophies—be these linked with religious, ideological, or socio-moral thought systems that are co-present in a society. The *double* feed-forward—"*meaning attaches to that which symbolises it*" guarantees that human psychological worlds become inevitably projected into the object world. If meaning X symbolizes object Y than Y becomes saturated with X (beyond the previous presentation), and feeds into further transformation of X, and so on. Here we see continuity of Baldwin's thought with Peirce's notion of the interpretant.

The critical case of negative aesthetic: the "ugly." Baldwin emphasized the negative aesthetic—"the ugly." It is a meaning that blocks the further growth of further aesthetic interest. There was an interesting resistance to the construction of meaning that this notion provided:

> We can play with anything; anything can be selected to take on the playful sort of semblance. But with art it is not so; the ugly is not beautiful, and nothing can make it so; it embodies materials *in which the sort of semblance operating in the aesthetic cannot go forward.* Some materials are always ineligible or unfit for aesthetic or artistic treatment. Let the term "ugly" to cover all cases of this—to include all objects which are positively unaesthetic—and inquire why they are so (Baldwin, 1911, p. 179, emphasis added).

Things and ideas that *cannot be idealized* are found "to be ugly." Yet it remains an interesting issue why they cannot be idealized—and what would happen if such affective-semiotic "block" is to be removed. Baldwin's thinking of the art in the 1908-1915 range was not confronted with the changes in European art movements after World War I. The case of an implanted urinal in the museum wall came into being only with the war (Marcel Duchamp—*The Fountain*—in 1917)—as the previously ugly becomes a social philosophy of beauty when the sounds of *Dada* could be heard in European art circles.

How can that happen? Baldwin points to the meaningful distancing of the work of art from the reality,

> The whole meaning must be uprooted from the sphere of the actual situations, individual or social, which may call for serious solution. The character in the play must be depicted as responsible to its requirements, not to ours; to his fellows, not to us. The requirements of his situation are set forth as if actual; *but as soon as these requirements are carried over into the world of actual personal obligation their aesthetic lesson is lost;* the envelope, the circumscribing semblance, is punctured. *Practical morals then succeeds beauty*; and the work of art gives place to the sermon of the homily (Baldwin, 1911, p. 198, added emphasis).

The play needs to be distant to be aesthetic—once that abstractive and affective distancing vanishes—so too would art. The artist may explore new meanings for one's own sake—yet in ways that are inter-subjectively triggering similar feelings of desired affective play-at-distance among others. This may come with delay—the personal and the social directions of feeling needs are not necessarily in synchrony. The history of initial failures of art to gain public attention—or to trigger public condemnation—only to become aesthetically pleasing for many years later—gives us ample evidence of the work of the mechanism Baldwin described (Schlicht and Hollein, 2008).

Striking the balance. The transition from the 1911 book to the present treatise was implicated by Baldwin's recognition of the omission of from the third volume, and the need to develop a third alternative—to the realism and phenomenalism split. This for him was *immediatism,*

> in which the opposition between real and phenomenal is annulled; in which thought and things come together as common content of contemplation—a direct presence, the awareness of which is itself just the apprehension of the real. This fully significant real, found in aesthetic contemplation, assumes the partial reals of knowledge and will and reconciles their oppositions; in this sense our view is realistic. But it brings the real, as most adequately apprehended, into a mode of experience; in this sense it is idealistic. Yet it abrogates all trans-experiential reference, since all controls, external and internal, are fused in one: in this sense it is *neither realistic nor idealistic* (Baldwin, 1911, p.271).

Maintaining the balance of the connected opposites—the credo Baldwin had learned from McCosh in his student days in Princeton—is maintained in this focus on the fusion of the opposites in the moment of immediate experiencing. Baldwin's immediatism in 1911 was not ready for a dialectical synthesis:

> this point of view seems to give philosophical standing to the resort to feeling, contemplation, intuition, immediacy—by whatever term the transcending and fusing of the motives of intellect and will be known—*not in the Hegelian sense of a solution which is always logically more abstract,* or in the sense of their mere reduction in the melting-pot of mystic feeling, but *in the simple empirical sense in which the movement of experience itself comes to unity and equilibrium in contemplation* (ibid, added emphasis).

Baldwin promised to cover the issues in the fourth volume—the present book—on aesthetic synthesis and its uses in religious contemplation (Baldwin, 1911, p. xiv).

§ 3. Selected Highlights of the *Genetic Theory of Reality*

Baldwin's theoretical contributions to the social sciences culminated in this book. All the sociogenetic roots of human personality that were the focus of his work in the 1890s were by the 1910s turned into an elaborate coverage of the person who creates one's own higher affective subjectivity—aesthetic relation with the surrounding social world—through that social world (rather than under one-sided influence of that world). The social world is the key resource for development of such higher-level subjectivity. Baldwin's initial trust in logic became modified by the end of his four-volume series. The focus on reality and his developmental method (see further—chapter 12 paragraph 2 in this book). This was reflected in the title—where *Reality* in 1915 replaces *Logic* of 1906-1911.

Genetic Theory of Reality includes three elaborations of three problems that are as central (which means—still unsolved) to our social sciences as they were at his time—the role of religious feelings in the personal lives of human beings, the unity of the personal and the social in the human *psyche* (nowadays addressed by the sociocultural perspective—Valsiner and Rosa, 2007), and the understanding of synthesis in psychological processes. In addition it includes an innovative coverage if the notion of logic—as Baldwin can be considered a forerunner of the deontic logic (Valsiner, 2009) where generic values play a central role in deductive calculations.

As the reader will immediately find out, *Genetic Theory of Reality* is a complicated book where references to other volumes of *Thought and Things* are made to provide the background for his newest elaborations. His own creation of ideas—often introducing neologisms—is a developmental process in which *Genetic Theory of Reality* is the point of final arrival—or exodus. After that book, Baldwin did not publish further theoretical contributions—even as he remained active as an intellectual.

Baldwin's writing style is complex—with contemplative returns to key ideas, allusions to the key notions from other authors ("blooming confusion" of William James, "*elan vital*" of Henri Bergson, chapter 10 paragraph 1) and small side stories. Both his contemporaries and later readers found him extensively verbose—yet that claim can probably be made of any philosopher. Baldwin was really a philosopher in biology and the social sciences—at a time when it was generally appreciated. In some ways, *Genetic Theory of Reality* is a contribution to the developmental science on its social and psychological side similar to Baldwin's co-elaboration of the Organic Selection Theory in 1896 for evolutionary biology. Both innovations start from the same source—the focus on persistent imitation, and the semiotic logic of Charles Sanders Peirce.

Development and Vanishing of Mediation: From Pre-Logical to Hyper-Logical

Peirce's semiotics was keenly present in Baldwin's *Dictionary of Philosophy and Psychology* (1901-1905), and in his thinking about the role of signs. Baldwin's focus was in the personal use of signs in the making of knowledge—the process of ejection—where the person imaginatively relates the given sign with the present context:

> The instrumental meaning is always and everywhere a re-reading imaginatively, purposefully, personally of an actual or truthful meaning, and the truthful reading is always and everywhere a re-reading as common, stereotyped, actual of an imaginative personal construction.... We make-believe in order that we may believe! (Baldwin, 1908b, pp. 183-184).

It is in that act of imagination that development of new meanings—mediation devices—and of the rules of their links (logic) emerges. Baldwin was not satisfied by the dichotomy of the philosophies of his time looking at the contrast of thinking logically between the "primitive" and "cultivated" man. The first was supposed to think a-logically, while the latter was presented—in an act of self-satisfaction by the "cultivated man" himself—as logical.

In contrast, Baldwin demonstrates how human cognitive functions are organized into a hierarchy of three—from "pre-logical" (i.e., what was seen as "primitive") to "logical" (mediated by signs, and rules of logic) to "hyper-logical" states. In both the "pre-logical" and "hyper-logical" states the canons of classical logic do not apply—the Law of Excluded Middle is not honored (chapter 4, paragraph 4, and chapters 14 and 15 in the present book). The canons of classical logic apply only in the middle state—the "logical state" and are transcended by "genetic logic" of development to the level of the "hyper-logical."

Religious Feeling within the Self

Baldwin was slow in bringing the coverage of religious phenomena into his theoretical system—he promised it in 1911 and delivered the promise in the present book (see chapters 6 and 7).

The notion of the person being social through the ejective processes was with Baldwin since the 1890s—as I pointed out above—and the coverage of religious feelings in this book sets these up in an elaborate framework. Ejection entails the person's turning to another person—or a sign—populating it with one's own meaning—and then treating that meaning as if it were a "message" from that external source for the person oneself.[12] The ejected material returns—being modified through the act—to the ejector. The ejection process operates through the uniting of the general and the unique—in ways that extend the duration of the unique and establish its new characteristics. Baldwin schematized that already in 1908 (after Baldwin, 1908a, p. 340):

A (([=becomes] A —— B

and the new structure AB incorporates C:

AB ((A —— B —— C

Conclusion:

A (([has become] ABC

Through the intermediary B, A has become C (as a part of the whole A-B-C). The whole entails inclusion (B includes A and other "A-like" members—e.g., a class membership) and generalization (C which includes B and A). The whole here surely is not equal to the sum of its part, but has an autonomous standing (ABC), while the parts in it are detectable. That scheme fits the development of religious feelings—a person by including oneself within a community (B) establishes a distant but total relation with a deity (C). This feature of ejective relating can help to understand Baldwin's claims in chapter 7 (paragraph 9) as deities operate through humanized intermediaries—which are created by the persons themselves as if coming from outside.

The Dynamic Field of Synthesis

Baldwin labeled this focus "pancalism"—"aesthetic immediatism" or "contemplationism" (see chapter 10 paragraph 1)—explaining it at the end of the third volume of *Thought and Things* (Baldwin, 1911, chapter 15), or "constructive affectivism," as it was claimed to unite aesthetic feeling with a distanced view of the object.

Aesthetic synthesis entails *retaining the object-subject differentiation* (i.e., the subject does not "fuse" himself with the object), yet it *simultaneously entails the emergence of a novel feeling that overwhelms the subject.* It is not a new abstract *concept*—distanced and "dry" (here Baldwin remains true to keeping distant from Hegel). It is a new *complex*—rather than a new concept. In that complex it is the affective hyper-generalization that leads the development of the complex—the notion "how *beautiful!*" (or "how *ugly!*") is a holistic personal feeling which overwhelms the subject who is relating with the object—a painting, a sculpture, the body of one's lover, a sunset, or a symbol of national identity.

In aesthetic experience, the singular event (a person's encounter with a particular art object in a here-and-now situation) becomes generalized by the person to represent something at the level of great abstraction. The aesthetically-operating person relates to the object in terms that are generic—even as any actual encounter is not different from a mundane one. We may pass by many people in a busy city not noticing any of them—and suddenly we find ourselves cherishing the beauty of a face of a person who just momentarily emerges from the crowd—vanishing at the next instant.

The person's arrival at the aesthetic synthesis is made possible by the process of tension between the aesthetic and mundane areas of reflection. That tension finds its solution in the hypergeneralized aesthetic judgment. The work of such tension in art was well described by Lev Vygotsky a decade after Baldwin (1925, in his *Psychology of Art*) as he discussed the ways in which an author of a story creates the conditions for such aesthetic synthesis:

> a work of art always contains a certain kind of contradiction, a certain inner incongruity between the material and the form. We can say that the author intentionally chooses a difficult, resisting material—of the kind that will resist by its inherent properties all his efforts to say what he wants to say. The more uncertain, resistant, and hostile the material, the more suitable it seems to be for the author. The form given by the author to this material is not intended to reveal the properties of the material itself.... On the contrary, its purpose is to overcome these properties... (Vygotsky, 1987, p. 156—English translation in Vygotsky, 1971, p. 160).

Creating a work of art is a dialogical struggle—of the Subject with the Object. Vygotsky went beyond Baldwin—explicitly embracing some part of Hegel's dialectics[13] while remaining true to Spinoza in his focus on affect. Baldwin continued on his McCoshian trail of maintaining a balance—yet giving it a tensions-maintaining form.

§ 4. *The Final Exile: England and France*

After the Mexico stay ended, Baldwin immigrated to Europe. He settled in England (Horley, 2001) and made frequent trips to France, where he remained intellectually and socially active until the end of his life (1934). Aside from the references to his continuing luncheons with Pierre Janet, history of psychology has failed to highlight his further contributions—which of course did not surpass the peak achieved with the *Genetic Theory of Reality*. Yet that may be only partially accurate, as his wider contributions were at the time visible:

> It is possible that the most significant aspect of Baldwin's life is that which can not be reconstructed from his printed works. As a mediating figure between European and American culture, he was one of the outstanding figures of his time. He became, for many, a sort of symbol of cooperative scholarship and science. When, after frequent sojourns in Paris, he finally took up his permanent residence there, this aspect of his activity increased in importance.... His own personal involvement, through the injuring of his daughter in the torpedoing of the *Sussex* [an ocean liner attacked in World War I], intensified his emotional participation, but his long list of national and political studies which were the product of that time, are all on the highest level and reflect the innate idealism of the man (Urban, 1942, p. 306).

Baldwin in the post-1915 years became a cultural interpreter—of the world of United States, France (and Continental Europe as a whole), and England. Much of it was linked with the World War I—the understanding by Europeans of the United States was as profoundly limited as was its opposite—Europe misunderstood by the Americans. The latter is no novelty—as it is deeply rooted on the exceptionalist calling that people raised in the United States often accept as a given.

In the context of the raging war Baldwin needed to distance his aesthetic synthesis—pancalism—from that of the over-personal suggestions of national identities that rage during wars. In a critique of Germany for the superimposition of over-personal (state) values, Baldwin's distance from Hegel in philosophy comes out clearly—he combines Nietzsche and Hegel as roots of the German state—"the Hegelian State" (Baldwin, 1916, p. 17). In general, his "war notes" (Baldwin, 1919) convey the anxieties of seeing all sides in conflict misunderstanding one another's cultural-historical agendas.

In his efforts to explain differences between societies, his personal attachment to France and critique of America were recurrent—and for good reasons, both personal and socio-political. In a sense, his guided immigration to Europe made him free—from the social context of the United States and from the competitive intrigues of his fellow American psychologists, many of them immigrants from Europe. In a bulletin written for the American Women's Club in Paris (dated 1923) he explains (to American women living in France) the life in French ways:

> The French individual ... minds his own business, at least ostensibly, and expects even more ostensibly that others mind theirs. *Ça, ce n'est pas mon affaire.* Hence it is that an extreme laissez-faire runs all through their social life. No one feels called upon to correct abuses, provided they do not affect his personal rights. They put up with antiquated methods of doing business and old-fashioned apparatus in daily usage. Hasn't the individual the right to work in the way and with the means of his fathers? (Baldwin, 1926, p. 141)

In contrast, Baldwin reminds American women that in America:

> The arm of the law reaches everywhere, repressing originality along with caprice, and producing a virtuous but uninteresting uniformity where in France there is an interesting though perhaps, on the surface, less edifying variety. Just now there is a recrudescence of this form of legalism in the United States, the daily papers reporting new regulations on things as diverse and trivial as bathing-suits, jaywalking, and smoking in public places (Baldwin, 1926, p. 143).

The contrast was consolidated into slogans—"Liberty guided by reason" (French) and "freedom armed with law" (America). His form of rationalism was—as he clearly understood himself (Baldwin, 1918, p. viii)—fitting in France where systematic theorizing and aesthetic sentiments were the natural gift of the cultural history.

§ 5. *The Man and His Word*

James Mark Baldwin was no simple man—ambitious, arrogant, and amiable at the same time. Only the very best of American intellectuals of his time—William James and Charles Sanders Peirce—could recognize and appreciate Baldwin as a thinker. Hundred years are gone since these days—and our needs to understand the complex nature of human affective processes is on the upsurge. Again psychologists, anthropologists, and educationalists talk of the centrality of the body in human cultural living. Again the nature of the complex affective phenomena becomes recognized as hard to approach—even through modern technologies. We may now have powerful tools—but few ways of thinking about the object of our investigations. A return to the efforts of James Mark Baldwin to make sense of affective synthesis is thus imminent—and this book is a living testimony to this.

Genetic Theory of Reality gives us a glimpse into one of the brilliant minds in the history of knowledge in America.

The context of Baldwin's ideas was different from ours. A hundred years ago there were few scholars—in contrast to our time where Ph.D.-degreed persons are almost mass produced—and these few usually worked in multiple fields of knowledge. Similarly to others—William James, Josiah Royce, John Dewey, George Herbert Mead, Charles Sanders Peirce—Baldwin was a complex figure in the academic world—a philosopher, an experimenter, a public citizen. In our contemporary world he is often presented as one of the key figures in developmental psychology, and as founder of psychology laboratories in different universities in North America. In evolutionary theory he is known as a co-founder of the organic selection perspective in evolutionary thought (but he was not its complete originator—see Bateson, this book). He was also a *connoisseur* of the pleasures of living, from art to other forms of sensuality, a dedicated parent to his children, editor of books and journals, administrator of emerging university departments and commentator on world politics. He was an ambitious man. Yet perhaps the most important feature of his life is that he was a creative thinker—one who did not follow fashions, and who tried—and tried again and again—to provide solutions to basic philosophical problems of the social nature of the deeply personal creature—the human being. He was—alongside George Herbert Mead—a leading sociogenetic theorist, the best that our New World has managed to produce. The implications of his major contribution—the logic of development that leads to aesthetic synthesis—are yet to be expanded and innovated. The present book is an introduction to the issues James Mark Baldwin raised—and persistently tried to resolve. The persistent imitation of Baldwin's ideas is a promising novelty for the future.

Acknowledgments

I am grateful to Brady Wagoner for reading and commenting upon a previous version of this manuscript, and to René van der Veer for many years of sharing the adventures into the unknown histories of good thinkers.

Notes

1. Michel Foucault (1998, p. 185) has captured this general social movement succinctly: "In civilizations without ships the dreams dry up, espionage takes the place of adventure, and the police that of the corsairs."
2. Dugald Stuart (1753-1828)—Scottish philosopher, student of Thomas-Reid.and Adam Ferguson, and author of textbook *Outlines of Moral Philosophy.*
3. Thomas Reid (1710-1796)—Scottich philosopher, founder of the Scottish perspective of common sense philosophy—common sense realism.
4. Friedrich Paulsen (1846-1908)—German philosopher, student of Gustav Fechner, who was philosophy professor in Berlin until his death, most American psychologists who visited Berlin in the last decades of nineteenth century encontered his seminars.
5. Hermann Lotze (1817-1881) was a German philosopher and logician whosae roots were in the philosophies of Schelling, Fichte, and Hegel. His *Systeme der Philosophie* (2 volumes—Logik, 1874, Mataphysik, 1879) were—aside from Wundt's writings—second major German source for Baldwin's ideas.
6. In the 1880s the Canadian provinces of the British Crown looked down upon the booming and buzzing industrializing confusion in the United States, which was "widely viewed in Ontario as being crude and vulgar. America's strong emphasis on individualism was seen as leading to, and even indulging, unseemly personal excesses.... British traditions would be overwhelmed by what they perceived to be the "unsavory" customs of their enormous southern neighbor" (Green, 2004, p. 131).
7. Jean-Baptiste Lamarck's *Philosophy of Zoology* was published in 1809—and led to recurrent denunciations of his basic idea of modi-

fiability of the organisms by experience. The popular minds made much out of the ridiculous image of the giraffe extending his or her neck by reaching up in the trees for food. What got lost was the idea of *limited but possible* modifiability of the nature—the idea that guides our present day biological explorations.

8. The English publisher insisted on the title *Thought and Things* for his main work—and Baldwin grudgingly accepted—with result of arriving at a remarkably long full title—which is: *Thought and Things: A Study of the Development and Meaning of Thought or Genetic Logic*. The French title was *Le Jugement et la Connaiscance: Logique fonctionelle*.
9. Baldwin claimed to build on Peirce's contributions to his *Dictionary of Philosophy and Psychology*—Baldwin, 1908a, p. ix.
10. In his glossary of terms—see at the end of this book—Baldwin describes the will-to-believe as "readiness to accept one alternative in preference to others, in the absence of logical proof or strong rational presumption."
11. The dedication reads "To the memory of William James whose teaching was always inspiring and *whose friendship was always true*" (Baldwin, 1911, added emphasis). The story of American psychologists' and university officials' political execution of Baldwin's position in American academia after public knowledge of his visit to a Baltimore brothel became known is a saga of betrayals and hushed-up institutional façade defenses by his colleagues in addition to university officials. William James was almost the only one of his colleagues who persistently tried to block that punishment for "crimes" never committed (see Richards, 1987, pp. 496-501).
12. This idea is in our times becoming influential in the social sciences through the discovery of the intellectual depth of the work of Mikhail Bakhtin:

> The word of language [*slovo iazyka*—in Russian]—is half alien [*chuzoye*—not belonging to me and unknown—in Russian] word. It becomes "one's own" when the speaker inhabits [*naselit*, in Russian] it with his intention, with his accent, masters [*ovladeet*, in Russian] the word, brings it to bear upon his meaningful and expressive strivings. Until that moment of appropriation [*prisvoenie* in Russian] the word is not existing in neutral and faceless language (as the speaker does not take the word from a dictionary!), but [it exists] on the lips of others, in alien contexts, in service of others' intentions: from here it has to be taken and made into one's own (Bakhtin, 1934/35—published in 1975, p. 106).

The speaker inhabits the word with intention—an example of ejection.

13. While others remained without acceptance—the notion of higher order abstract notion arrived at via dialectical leap. Baldwin's synthesis is that of abstracted field as it becomes generalized—and not of abstract point-like general concepts emerging from the field.

References

Albertazzi, L., Jacquette, D., and Poli, R. (Eds.) (2001). *The school of Alexius Meinong.* Aldershot: Ashgate.

Bakhtin, M. M. (1934/1975). Slovo v romane [Discourse in the novel]. In M. Bakhtin, *Voprosy literatury i estetiki* (pp. 73-232). Moscow: Khudozhestvennaya Literatura. [in English translation Bakhtin, 1981]

Bakhtin, M. M. (1981). *The dialogic imagination.* Austin, Tx.: University of Texas Press.

Baldwin, J. M. (1892a). Origin of volition in childhood. *Science, 20* (No. 511), 286-287.

Baldwin, J. M. (1892b). Among the psychologists of Paris. *The Nation, 55,* No. 1413, 68.

Baldwin, J. M. (1892c). With Bernheim at Nancy. *The Nation, 55,* No. 1415, 101-103.

Baldwin, J. M. (1894). Imitation: A chapter in the natural history of consciousness. *Mind, 3* (new series), 26-55.

Baldwin, J. M. (1895). *Mental development in the child and the race.* New York: Macmillan.

Baldwin, J. M. (1897). *Social and ethical interpretations in mental development.* New York: MacMillan.

Baldwin, J. M. (1898). On selective thinking. *Psychological Review, 5,* 1, 1-24.

Baldwin, J. M. (1902a). *Fragments in philosophy and science.* New York: Charles Scribner's Sons.

Baldwin, J. M. (1902b). *Development and evolution.* London: MacMillan.

Baldwin, J. M. (1905). Sketch of the history of psychology. *Psychological Review. 12,* 144-165.

Baldwin, J. M. (1906). *Thought and things: A study of the development and meaning of thought, or genetic logic.* Vol. 1. *Functional logic, or genetic theory of knowledge.* London: Swan Sonnenschein & Co.

Baldwin, J. M. (1908a). *Thought and things: A study of the development and meaning of thought, or genetic logic.* Vol. 2. *Experimental logic, or genetic theory of thought.* London: Swan Sonnenschein & Co.

Baldwin, J. M. (1908b). Knowledge and imagination. *Psychological Review, 15,* 181-196.

Baldwin, J. M. (1911). *Thought and things: A study of the development and meaning of thought, or genetic logic.* Vol 3. *Interest and art being real logic.* London: Swan Sonnenschein & Co.

Baldwin, J. M. (1916). *The super-state and the 'eternal values'.* London: Oxford University Press.

Baldwin, J. M. (1918). Avant-propos de l'édition Française. In J. M. Baldwin, *Théorie génétique de la réalité. Le pancalisme* (pp. vii-viii) Paris: Félix Alcan.

Baldwin, J. M. (1919). *Paroles de guerre d'un Americain 1914-1918.* Paris: Felix Alcan.

Baldwin, J. M. (1926). *Between two wars 1861-1921.* Vol. 1-2. Boston, Ma.: The Stratford Company.

Baldwin, J. M. (1930). James Mark Baldwin. In C. Murchison (Ed.), *A history of psychology in autobiography.* Vol. 1 (pp. 1-30). New York: Russell & Russell.

Brownson, O. (1842). Schmucker's psychology. *United States Magazine and Democratic Review,* 11, 358. [cited via Rieber, 1998]

Cairns, R. B., Elder, G., and Costello, E. J. (Eds.) (1996). *Developmental science.* New York: Cambridge University Press.

Chang, R. S. (Ed) (2009). *Relating to environments: A new look at Umwelt.* Charlotte, N.C.: Information Age Publications.

Dewey, J. (1896). The reflex arc concept in psychology. *Psychological Review, 3,* 3, 357-370.

Foucault, M. (1998). Different spaces. In J. Faubion (Ed.), *Essential works of Michel Foucault 1954-1984.* Vol. 2 (pp. 175-185). New York: The New Press.

Gallegos, X (1980). James M. Baldwin's visits to Mexico. *American Psychologist, 35,* 8, 772-773.

Giddings, F. H. (1899).The psychology of society. *Science, 9,* 210,16-23.

Gottlieb, G. (1992). *Individual development & evolution: The genesis of novel behavior.* New York: Oxford University Press.

Gottlieb, G. (1997). *Synthesizing nature/nurture.* Mahwah, N.J.: Erlbaum

Green, C. (2004). The hiring of James Mark Baldwin and James Gibson Hume at the University of Toronto in 1889. *History of Psychology, 7,* 2, 130-153.

Grossmann, R. (1974). *Meinong.* London: Routledge & Kegan Paul.

Horley, J. (2001) After the Baltimore affair: James Mark Baldwin's life and work, 1908-1934. *History of Psychology, 4,* 1, 24-33.

Kull, K. (2001). Jakob von Uexküll: An introduction. *Semiotica,* 134, 1-59.

Maier, B. N. (2005). *The separation of psychology and theology at Princeton, 1868-1903.* New York: Mellen Press.

Moscovici, S. (1976). *Society against nature: The emergence of human societies.* Atlantic Highlands, N.J.: Humanities Press.

Schlicht, E, and Hollein, M (Eds.) (2008). *René Magritte 1948: La période vache.* Frankfurt-am-Main: Schirn Kunsthalle.

Schmit, D. (2005). Re-visioning antebellum American psychology: The dissemination of mesmerism 1836-1854. *History of Psychology, 8*, 4, 403-434.

Ribot, T. (1886). *German psychology of to-day.* New York: C. Scribner's sons.

Richards, R. J. (1987). *Darwin and the emergence of evolutionary theories of mind and behavior.* Chicago: University of Chicago Press.

Rieber, R. J. (1998). Americanization of psychology before William James. In R. J. Rieber and K. Salzinger (Eds), *Psychology: Theoretical-historical perspectives* (pp. 191-215). 2nd ed. Washington, D.C.: American Psychological Association.

Simpson, G. G. (1953). The Baldwin Effect. *Evolution, 7*, 110-117.

Spilka, B. (1986). Religion and science in early American psychology. *Newsletter of APA Division 36 (Psychologists interested in religious issues)*, 11, 3, 2-4.

Urban, W. (1942). James Mark Baldwin- Co-Editor, *Psychological Review*, 1894-1909. *Psychological Review, 42*, 303-306.

Valsiner, J. (2009). Baldwin's quest: A universal logic of development. In J. W. Clegg (Ed.), *The observation of human systems: Lessons from the history of anti-reductionistic empirical psychology* (pp.45-82). New Brunswick, N.J.: Transaction Publishers.

Valsiner, J., & van der Veer, R. (2000). *The social mind: Construction of the idea.* New York: Cambridge University Press

Valsiner, J., & Rosa, A. (Eds.) (2007). *The Cambridge Handbook of Sociocultural Psychology.* New York: Cambridge University Press

Valsiner, J., and Rudolph, L. (2008). Who shall survive? Psychology that replaces quantification with qualitative mathematics. Paper presented at the *29th International Congress of Psychology, Berlin*, July, 21 in the framework of the Symposium Why Psychology Moves towards the Qualitative Epistemological Foundations (Günter Mey and Jaan Valsiner, co-conveners)

von Uexküll, J. (1926). *Theoretical Biology.* New York: Harcourt, Brace & Company, Inc.

Vygotsky, L. S. (1987). *Psikhologia iskusstva.* Moscow: Pedagogika (in Russian)

Vygotsky, L. S. (1971). *The psychology of art.* Cambridge, Ma.: MIT Press.

Weber, B. H., and Depew, D. J. (Eds.) (2003). *Evolution and learning: The Baldwin Effect reconsidered.* Cambridge, Ma.: MIT Press.

Wozniak, R. H. (1998). Thought and things: James Mark Baldwin and the biosocial origins of mind. In R. W. Rieber and K. D. Salzinger (Eds.), *Psychology: Theoretical-historical perspectives* (pp. 429-453). 2nd ed. Washington, D.C.: American Psychological Association.

SUPPLEMENTAL COMMENTARY

BALDWIN AND THE ADAPTABILITY DRIVER

Patrick Bateson[1]

FOR many years I have kept notebooks in which an uncoordinated collection of notes on interesting books, references to check, shopping lists, and other jobs to be done are stored. Every so often, when I think that I have had a good idea, I jot it down. What is startling is how often the very same idea that came to me with such freshness one morning, was exactly the same idea I jotted down a year before with the same sense of originality. What is more embarrassing is that when I go back far enough in my notebooks, I may discover that I recorded that very same idea at someone else's lecture. I am sure that I am not alone in this and many of the simultaneous "discoveries" must have stemmed from a common and often unacknowledged source. None of this implies that original thought does not exist, nor that failure to cite the first person to have the idea is a sign of poor scholarship or, worse still, intellectual theft. However, the unconscious processes that lie behind what the author supposes was a genuine act of creation are relevant to consideration of what is now known as the Baldwin effect.

§ *I. Organic Selection*

Modern thinking about the importance of behavioral plasticity in evolution is usually thought to stem from Baldwin (1896), but Lloyd Morgan (1896) and Osborn (1896) independently developed ideas about "organic selection," as the subject was called at the time. Lloyd Morgan's (1896) account of the process was particularly clear and may be paraphrased as follows:

(a) Suppose that a group of organisms that are capable of change in their own lifetimes are exposed to new environmental conditions.
(b) Those whose ability to change is equal to the occasion survive. They are modified. Those whose ability is not equal to the occasion are eliminated.
(c) The modification takes place generation after generation in the changed environmental conditions, but the modification is not inherited. The effects of modification are not transmitted through the genes.
(d) Any variation in the ease of expression of the modified character which is due to genetic differences is liable to act in favor of those individuals that express the character most readily.
(e) As a consequence, an inherited predisposition to express the modifications in question will tend to evolve. The longer the evolutionary process continues, the more marked will be such a predisposition.
(f) Thus plastic modification within individuals might lead the process and a change in genes that influence the character would follow; the one paves the way for the other.

It is obvious from this outline of the proposed process, that Lloyd Morgan was not suggesting genetic inheritance of acquired characters as a mechanism. Nor, less obviously and less consistently, were Osborn or Baldwin. The crucial postulate is a cost of operating the original process of phenotypic adaptation, a cost that can be subsequently reduced by genotypic change enabling Darwinian evolution to occur. Adaptability to new conditions might be physiological, such as coping with high altitudes by enhancing the oxygen carrying capacity of the blood. Lloyd Morgan's clear exposition gives the lie to Ancel's (1999) claim that it was Simpson (1953) who Darwinized organic selection.

Alister Hardy (1965) did more than most in intervening years to pick up on this point and stress that the process could be of great significance. He put it as follows:

> If a population of animals should change their habits (no doubt often on account of changes in their surroundings such as food supply, breeding sites, etc., but also sometimes due to their exploratory curiosity discovering new ways of life, such as new sources of food or new methods of exploitation) then, sooner or later, variations in the gene complex will turn up in the population to produce small alterations in the animal's structure which will make them more efficient in relation to their new behavior pattern; these more efficient individuals will tend to survive rather than the less efficient, and so the composition of the population will gradually change. This evolutionary change is one caused initially by a change in behavior (Hardy, 1965, p. 170).

Hardy envisaged a cascade of changes flowing from the initial behavioral event. Even without structural change, control of behavioral development might alter over time. A requirement for this to happen is that adapting to the new conditions be more costly than doing it easily without active modification. One instance might involve differential responsiveness to particular types of food. Many cases of choice of an unusual food for a given species are probably not due to genetic changes, but to the functioning of normal mechanisms in unusual circumstances. A group of animals might be forced into living in an unusual place after losing their way, but they cope by changing their preferences to suitable foods that are locally abundant. Later, those descendants that didn't need to learn so much when foraging might be more likely to survive than those that could only show a fully functional phenotype by learning. A cost was incurred in the time taken to learn. As a consequence, what started as a purely phenotypic difference between animals of the same species living in different habitats became a genotypic difference.

West-Eberhard (2003) cites Gause (1942) as offering the first example of organic selection operating in unicellular *Paramecium*. Another early example of how the process might work was provided by the great child psychologist Jean Piaget who began his career as a biologist and was much influenced by Baldwin. He studied pond snails in different Swiss lakes that differed in how much wave action occurred; however, his work did not come out in book form until the end of his life (Piaget, 1979). When he transferred snails from one lake to another, the morphology of the shell was changed in response to the musculature required for the wave action of that lake. The descendants of the transferred snails eventually inherited the changed morphology without needing to be exposed to the wave action of that lake. In the terms used by Lloyd Morgan, the initial change could involve adaptability by the individual snail; the adaptability is won at some cost so that descendants expressing the character more efficiently would be more likely to survive. Piaget believed that he had found a clear example of the inheritance of acquired characteristics and never seemed to have realized how his supposed Lamarckian effect could be Darwinized.

It is easy to see why the confusion might have persisted since organic selection was frequently given a Lamarckian spin, notably by Baldwin himself. Baldwin wanted to claim priority for the proposal, which had also been linked with Lloyd Morgan and Osborn. Baldwin argued that he had come up with the idea in a book published before the papers of the other two (Baldwin, 1895).

§ *2.Spalding's Contribution*

Griffiths (2003) noted that the book in which Baldwin claimed to have preceded Lloyd Morgan and Osborn does not state the organic selection principle. It merely states the individual's capacity to accommodate to a changing world and new behavior patterns so formed might then be transmitted from one generation to the next by social means. So if Baldwin's claim for priority had been accepted, it would have been for a proposal that did not include the second crucial phase of organic selection—namely the evolution of behavior that is expressed at lower cost than the original learned form.

Baldwin had published a different idea in 1895 but, even if it had not been different, it would not have established the precedence that Baldwin so much wished to have claimed for himself. For 22 years earlier a brilliant scientist called Douglas Spalding had published the true precursor of what came to be known as organic selection.

Spalding's article, republished in 1953, is famous largely because it provides the first clear account of behavioral imprinting with which Lorenz (1935) is typically associated. However, the article also set out the principle by which a learned pattern of behavior could, in time, be expressed spontaneously by individuals in subsequent generations. Since Spalding has not received as much credit as was due to him for his evolutionary ideas, it is worth quoting him at length.

> In seeking to understand the phenomena of instinct we of course get the full benefit of the law of Natural Selection, which, though it throws no light on the origin of anything, mental or physical-for, as Mr. Darwin says, it "has no relation whatever to the primary cause of any modification of structure" nevertheless helps us to understand the existence of instincts far removed from the circumstances or conditions of life under which they could have been acquired. Suppose a Robinson Crusoe to take, soon after his landing, a couple of parrots, and to teach them to say in very good English, "How do you do, sir?"—that the young of these birds are also taught by Mr. Crusoe and their parents to say, "How do you do, sir?"—and that Mr. Crusoe, having little else to do, sets to work to prove the doctrine of Inherited Association by direct experiment. He continues his teaching, and every year breeds from the birds of the last and previous years that say "How do you do, sir?" most frequently and with the best accent. After a sufficient number of generations his young parrots, continually hearing their parents and a hundred other birds saying "How do you do, sir?" begin to repeat these words so soon that an experiment is needed to decide whether it is by instinct or imitation; and perhaps it is part of both. Eventually, however, the instinct is established. And though now Mr. Crusoe dies, and leaves no record of his work, the instinct will not die, not for a long time at least; and if the parrots themselves have acquired a taste for good English the best speakers will be sexually selected, and the instinct will certainly endure to astonish and perplex mankind, though in truth we may as well wonder at the crowing of the cock or the song of the skylark. Again, turkeys have an instinctive art of catching flies, which, it is manifest, the creatures in their present shape may have acquired by experience. But suppose the circumstances of their life to change; flies steadily become more abundant, and other kinds of food scarcer: the best fly-catchers are now the fittest to live, and each generation they are naturally selected. This process goes on, experience probably adding to the instinct in ways that we need to attempt to conceive, until a variety or species is produced that feeds on flies alone. To look at, this new bird will differ considerably from its turkey ancestors; for change in food and in habits of life will have affected its physical conformation, and every useful modification

> of structure will have been preserved by natural selection. My point however is, that thus, by no inconceivable steps, would be produced a race of birds depending for all their food on an instinctive art, which they, as then constituted, could never have acquired, because they never could have existed without it.

Spalding's article was published in *Macmillan's Magazine* which was later renamed *Nature*. It was not an obscure publication and it seems inconceivable that the three authors of what was known later as "organic selection" had not read it. As is already obvious, I am entirely sympathetic to the thought that they unconsciously assimilated Spalding's ideas and then forgot where they came from. Be that as it may, the proposals were not taken seriously by mainstream evolutionary biologists for many years.

§ 3. *Relevance to Evolutionary Biology*

The great paleontologist, George Gaylord Simpson (1953), disputed the importance in biological evolution of what he called "the Baldwin effect" and most neo-Darwinists accepted his critique. To this day considerable skepticism remains about the universality or even the plausibility of the so-called Baldwin effect (see Weber & Depew, 2003).

Simpson believed that if a new phenotype were valuable to the organism, it would evolve along traditional Darwinian lines. Secondly, he argued that, if plasticity were a pre-requisite for the evolutionary process and was generally beneficial, it would be disadvantageous to get rid of it.

On Simpson's first point, if learning (as an example of one form of adaptability in animals) involves several sub-processes or steps as in operant chaining, then the chances against an unlearned equivalent appearing in one step in the course of evolution are very small (Bateson, 2004). However, with the learned phenotype as the standard, every small step that cuts out some of the plasticity with a simultaneous increase in efficiency is an improvement. As an example, consider the Galapagos woodpecker finch, which pokes sharp cactus spines into holes thereby obtaining insect larvae as food. Suppose it does so without much learning but that an ancestor did so by trial and error. In the first stage of the evolutionary process, a naïve variant of the ancestral finch, when in foraging mode, might, for example, have been more inclined to pick up cactus spines than other birds. This habit spread in the population by Darwinian evolution because those behaving in this fashion obtained food more quickly. At this stage the birds still learned the second part of the sequence. The second step would have been that a naïve new variant, when in foraging mode, was more inclined to poke cactus spines into holes. Again this second habit spread in the population by Darwinian evolution. The end result is a finch that uses a tool without having to learn how to do so. Simultaneous mutations increasing the probability of two quite distinct acts (taking cactus spines and poking them into holes in the case of the woodpecker finch) would be very unlikely. Learning makes it possible for them to occur at different times and in any order. Without learning processes, having one act but not the other has no value. As a matter of great interest, it seems to be the case that the Woodpecker finch is half down the evolutionary road from fully learned to fully spontaneous (Tebbich et al., 2001).

As far as learning is concerned, Simpson's second point of criticism is based on an inadequate understanding of how behavior is changed and controlled (Bateson, 2004). Learning in complex organisms consists of a series of subprocesses. If an array of feature of detectors are linked directly to an array of executive mechanisms as well as indirectly through an intermediate layer and all connections are plastic, then a particular feature detector can become non-plastically linked to an executive system in the course of evolution without any further loss of plasticity. Whether these replies to Simpson's objections can be applied as cogently to plants or less complex animals requires arguments that have, as far as I know, not yet been developed.

Nevertheless, Simpson's critique is much less cogent than the mainstream thinkers in evolutionary biology had supposed and many writers have been continued to be intrigued by the thought that some of the complex patterns of behavior expressed spontaneously by organisms might have been driven by learning at an earlier stage of evolution (e.g., Avital & Jablonka, Bateson, 1982, 1988; Gottlieb, 1992; Hardy, 1965; Tierney, 1986; Wcislo, 1988; Vancassel, 1990; West-Eberhard, 2003; Wyles et al., 1983).

High mobility by animals, such as that involved in active exploration or migration, would have frequently placed them in conditions that revealed heritable variation not previously apparent in the population. For that reason, behavior, along with other forms of dispersion, was likely to be important in initiating evolutionary change. In addition, behavioral adaptability of the animals would have helped buffer them against extinction in new conditions.

Consequently, interplay between this process and the expression of new variation could have readily occurred. Similarly, both processes might have been enhanced if the animals were capable of choosing the local conditions to which they were most suited and if they could actively alter their environment.

All the forms of behavior involved in such interplay require complex neural processing. So it would not be surprising if the rate of behaviorally induced evolutionary change were related to neural capacity and be a special feature of complex animals such as birds and mammals. The size of the brain relative to the rest of the body is not a perfect measure of neural capacity, but many would argue that it is a surprisingly good one. Presumably brain size relative to body size rather than absolute brain size is the better measure because animals with larger bodies have more sense organs to monitor than those with smaller bodies, they have more muscles to control and the general house-keeping jobs of maintaining the larger body are greater.

Assuming that relative brain size is a measure of the animal's ability to change its behavior in ways that might increase the rate of evolution, does evidence for such a relationship exist? Wyles et al. (1983) found the taxonomic groups that were evolving most rapidly were also the groups with the biggest brains relative to body size. Argument rages about the way in which they obtained their measures and what the measures mean. More recently, however, others have obtained correlations between behavioral innovation and brain size in both birds (Sol et al., 2005) and primates (Reader & Laland, 2002). Possibly the correlations are due to a third and unmeasured variable.

Nonetheless, the appearance of such evidence is welcome. It shows that workers in this field are not necessarily stuck with producing plausible scenarios. They can start to test possibilities by the use of comparative evidence.

§ 4. *Theoretical Models*

Numerous attempts have been made to model the so-called Baldwin effect. Some have used analytical population genetics models (e.g., Ancel, 1999). These suffer from the point that worried Simpson, namely that in a simple analytical model, any loss of plasticity carries a substantial cost. Hinton & Nowlan (1987) pioneered the use of a computational approach and many others have followed them (e.g., Red'ko et al., 2005). In most of these simulations, agents were selected for their ability to find a particular high-fitness phenotype in a single-peaked landscape surrounded by areas of lower adaptiveness, but Wiles et al. (2001) simulated the adaptability driver in a multi-peaked landscape using hierarchical modular tasks.

The metaphor of a fitness landscape, first proposed by Sewall Wright (1930), has been criticized by Gavrilets (2004). He suggested that the vast majority of biologically plausible topologies are not represented by three-dimensional peaks separated by valleys. With more than three dimensions affecting fitness, the topology is much more complex. Gavrilets' important argument radically shifts the key problem from the metaphorical crossing of valleys between peaks to choosing the most beneficial path in the sea of nearly neutral alternatives. When modeling the adaptability driver, using the reformulated framework of

Gavrilets, the problem is how to find the exact combination of a small number of mutations required that will generate the same phenotype as that obtained by learning. In collaboration with two Russian colleagues, Mikhail Burtsev and Konstantin Anokhin, I have been looking into ways in which this problem can be overcome.

A central component of our approach is that of a functional system theory developed by P.K. Anokhin (1974). Functional systems are dynamically self-organized collectives of elements that combine to achieve an adaptive result for the organism. Elements are necessary but not sufficient for the development of a functional system. They can be changed by single point mutations or completed by learning.

At an early stage in development, functional systems are concerned with general requirements of the organism such as the metabolism of each cell. Later in development the systems have more specific roles in the overall organization of the organism and involve higher levels of organization such as the coordination of an organ or, higher still, a sequence of different behavioral acts. Each functional system depends on the development of a prior system except at the most elementary level and as the process of building on the prior system proceeds, the capacity for solving increasingly complex challenges becomes greater. In the course of coping with a challenge, new functional systems may be formed as a means of achieving results that cannot be obtained by an already established functional systems.

Our model proposes that the development of novel adaptive systems may be added to the end of spontaneously unfolding developmental processes. When this construction involves learning, the new systems can subsequently join those processes that do not require learning by piece-meal mutations of component elements of the system in the course of evolution. The model simulates the way in which learning increases the rate of evolution. As the complexity of challenges from the environment become greater, so learning exerts an ever more powerful role in meeting those challenges and in opening up new avenues for subsequent evolution (Burtsev et al. in preparation).

§ 5. *The Verdict*

Given Spalding's (1873) precedence and the simultaneous appearance in 1896 of the ideas about "organic selection," it seems inappropriate to term the evolutionary process the "Baldwin effect." The trouble is that calling the proposed process the "Spalding effect" is not descriptive of what initiates the hypothetical evolutionary process. The notion of a behavioral driver in evolution was introduced by Wyles et al. (1983), but they laid primary emphasis on imitation and, of course, other active behavioral processes such as mate choice had already been recognized by Darwin (1871). I have a strong preference for a term that captures the all the active roles of the organism in the evolutionary process. That is how I come to the term "adaptability driver" (Bateson, 2005).

Adaptability operates immediately on the occurrence of a mismatch between an organism's phenotype and its environment. Most individuals will probably possess such capacity and survive, but if acquiring the appropriate behavior requires time and energy, then Darwinian evolution would tend to generate the appropriate response at less cost In the case of behavior the evolutionary outcome would involve a developmental process that did not require learning. The evidence that gave rise to Waddington's concept of "genetic assimilation" suggests a different evolutionary process (e.g., Waddington, 1953). Here the developmental response to a change in the environment is haphazard, many individuals may not survive and the subsequent driver of evolutionary change is not usually explained. For these reasons, I think it is useful to distinguish conceptually between the adaptability driver and genetic assimilation (Bateson, 1988, 2004). Nevertheless, developmental plasticity in all its forms is an area where phenotypic variation driven by environmental change can provide material for subsequent evolutionary change—as West-Eberhard (2003) has argued at length.

The Darwinizing of phenomena that look superficially Lamarckian was a useful contribution, but attention as been focused unduly on the replacement of the outcome of one developmental mechanism by another that generates the same outcome. First, ontogenetic adaptability may allow a lineage to occupy a new place that can then be exploited by entering many new niches within that new habitat. The Galapagos finches are a fine example (Grant, 1986). Second, opening up a new developmental pathway as a result of an environmental change does not necessarily lead to an outcome that remains stable in the face of a return to previous environment. Waddington's (1953) fruit fly experiment is just one illustration of innumerable possible scenarios. The developmental breakout may provide radically new opportunities for those individuals equipped with the new phenotype. As

West-Eberhard (2003) has noted, this is where Baldwin genuinely opened up a different debate from one that centered narrowly round organic selection. For that reason we should forgive his self-aggrandizement. As far as he incorporated without acknowledgement the ideas of others into his own grand schemes, there is nothing to forgive. We all do it.

Note

1. Patrick Bateson is in the Sub-Department of Animal Behaviour at the University of Cambridge (UK).

References

Ancel, L.W. (1999) A quantitative model of the Simpson-Baldwin effect. *J. Theor. Biol.* 196, 197-209.

Anokhin, P.K. *Biology and neurophysiology of the conditioned reflex and its role in adaptive behavior.* (Pergamon Press, Oxford, 1974).

Avital, E. & Jablonka, E. (2000) *Animal traditions: behavioural inheritance in evolution.* Cambridge: Cambridge University Press.

Baldwin, J.M. (1895) *Mental development in the race and the child: methods and processes.* London: Macmillan.

Baldwin, J.M. (1896) A new factor in evolution. *Amer. Nat.*, 30, 441-451, 536-553.

Bateson, P. Behavioural development and evolutionary processes. In: King's College Sociobiology Group (eds.) *Current Problems in Sociobiology.* pp. 133-151. Cambridge: Cambridge University Press.

Bateson, P. (1988) The active role of behaviour in evolution. In Ho, M.-W. & Fox, S. (eds.) *Process and Metaphors in Evolution,* pp. 191-207. Chichester: Wiley.

Bateson, P. (2004) The active role of behaviour in evolution. *Biol. Philos.* 19, 283-298.

Bateson, P. (2005) The return of the whole organism. *J. Biosci.*, 30. 31-39.

Bateson, P. (2006) The Adaptability Driver: links between behaviour and evolution. *Biol. Theory: Integrating Development, Evolution, and Cognition,* 1, 342-345 (2006).

Burtsev, M.; Anokhin, K.; Bateson, P. Learning can guide the evolution of complex biological systems. In prep.

Darwin C. (1871) *The Descent of Man, and Selection in relation to Sex.* London: Murray.

Gause, G.F. (1942) The relation of adaptability to adaptation. *Q. Rev. Biol.,* 17, 99-114.

Gavrilets, S. (2004) *Fitness landscapes and the origin of species.* Princeton, N.J.: Princeton University Press.

Gottlieb, G. (1992) *Individual Development and Evolution.* New York: Oxford University Press.

Grant, P.R. (1986) *Ecology and Evolution of Darwin's Finches.* Princeton, NJ: Princeton University Press.

Griffiths, P.E. (2003) Beyond the Baldwin effect: James Mark Baldwin's "Social Heredity," epigenetic inheritance, and niche construction. In Weber, B. H., & Depew, D. J., eds. *Evolution and learning: the Baldwin effect reconsidered.* pp.193-215. Cambridge, MA: MIT Press.

Hardy, A. (1965) *The Living Stream.* London: Collins.

Lloyd Morgan, C. (1896) On modification and variation. *Science,* 4, 733-740.

Hinton, G.E. & Nowlan, S.J. How learning can guide evolution. *Compl. Syst.* 1, 495-502 (1987).

Lorenz, K. (1935) Der Kumpan in der Umwelt des Vogels. *J. Ornithol.* 83, 137-213, 289-413.

Osborn, H.F. (1896) Ontogenic and phylogenic variation. *Science* 4: 786-789.

Piaget, J. (1979) *Behaviour and evolution.* London: Routledge & Kegan Paul.

Reader, S.M. & Laland, K.N. (2002) Social intelligence, innovation, and enhanced brain size in primates. *Proc. Natl. Acad. Sci.* 99, 4436-41 (2002).

Red'ko, V.G.; Mosalov, O.P. & Prokhorov, D.V. (2005) A model of evolution and learning. *Neural Net.* 18, 738-745 .

Simpson, G.G. (1953) The Baldwin effect. *Evolution* 7, 110-117.

Sol, D.; Lefebvre, L. & Rodriguez-Teijeiro, J.D. (2005) Brain size, innovative propensity and migratory behaviour in temperate Palaearctic birds. *Proc. R. Soc. B.,* 272, 1433–1441.

Spalding D. (1873) Instinct with original observations on young animals. *Macmillan's Magazine* 27: 282-293.

Tebbich, S.; Taborsky, M.; Fessl, B. & Blomquist, D. (2001) Do woodpecker finches acquire tool-use by social learning? *Proc Roy. Soc. B,* 268, 2189-2193.

Tierney, A.J. (1986) The evolution of learned and innate behavior: contributions from genetics and neurobiology to a theory of behavioural evolution. *Anim. Learn. Behav.,* 14, 339-348.

Vancassel, M. (1990) Behavioral-development and adaptation—an assimilation of some of Waddington ideas. *Behav. Proc.,* 22, 23-31.

Waddington, C.H. (1953) Genetic assimilation of an acquired character. *Evolution*, 7, 118-126.

Wcislo, W.T. (1989) Behavioral environments and evolutionary change. Ann. Rev. Ecol. Syst., 20, 137-169.

Weber, B.H., & Depew, D.J., eds. (2003) *Evolution and learning: the Baldwin effect reconsidered.* Cambridge, MA: MIT Press.

West-Eberhard, M.J. (2003) *Developmental Plasticity and Evolution.* New York: Oxford University Press.

Wright, S. 1930. Evolution in Mendelian populations. *Genetics*, 16, 97-159.

Wyles, J.S.; Kunkel, J.G. & Wilson, A.C. Birds, Behavior, and Anatomical Evolution. *Proc. Nat. Acad. Sci.* 80, 4394-439 (1983).

PREFACE

IN this book the matter appears which I had intended to place in the projected fourth volume of the work *Thought and Things or Genetic Logic.*[1] The change of plan has resulted from two considerations. The first and less important reason for it is that the London publishers of the latter work having gone into the hands of a "receiver," there promised to be delay and uncertainty as to the appearance of the new volume.[2] The more essential consideration, however, is that by making of this volume a separate work, I am able to include in it anthropological and historical matter which I had promised to the present publishers some time before. In fulfilling that promise, I am therefore completing the scheme of treatment of Genetic Logic, as announced in the Introduction to the first volume of *Thought and Things*, and at the same time giving to the topic of Genetic Morphology (the last division of Genetic Logic) more extended treatment and wider scope. I find, in fact, of great interest the

[1] London, Geo. Allen & Co. (now Allen & Unwin); New York, Macmillan. vol. i., "Functional Logic," 1906; vol. ii., "Experimental Logic," 1908; vol. iii., "Interest and Art, Genetic Epistemology," 1911.

[2] It is only just that I should add that the acknowledgments made to the English publishers, in the Preface to volume iii., were intended for Messrs. Swan, Sonnenschein & Co., whose business was taken over—with what success I have indicated above—by Messrs. Geo. Allen & Co.

confirmations coming to the theory of Interpretation—as issuing from the analyses of the larger work—from the treatment of the racial and historical problems (in Part II. of the present text).[1]

The close relation subsisting between this volume and the others mentioned is also seen in the frequent allusions to the latter found in the present text. By means of such references many repetitions are avoided; and at the same time those detailed analyses are summarised upon which rest the conclusions of this work. This is also, of course, the excuse for such citations: they are the references necessarily made from one part of a literary whole to other parts.

The main conclusion itself, the doctrine of Pancalism, is the author's matured contribution to the theory of reality—the philosophy to which his professional life of study, teaching, and writing has led him. It is now for the first time stated with sufficient fulness and with historical perspective. It was, however, thought out some time ago, and the principal point has had more or less definite statement in print on more than one occasion. Besides the exposition contained in vol. iii. of *Thought and Things*, chap. xv. (see also the Preface to the same volume, where the term Pancalism is first made use of), brief statements are to be found in articles in the *Psychological Bulletin*, vol. iv., April 15, 1907, and (especially) the *Psychological Review*, May, 1903, and May, 1908; in *Darwin and the Humanities*, London ed., 1910, p. 93; and in the Preface to vol i. of *Thought and Things*, 1906, p. x, to which the motto

[1] Confirmations briefly anticipated in part, in the Preface to volume iii., the final chapter of which also anticipates, in the form of a programme, the general theory of Pancalism now fully worked out.

τὸ καλὸν πᾶν was affixed as an indication of the result to which the entire research had led the author. Further, apart from lectures delivered in Princeton and Baltimore[1] since about 1900, an early printed statement, which it is now relevant to recall, is that made in the Preface, p. ix, to the volume of collected essays entitled *Fragments in Philosophy and Science* (New York, Scribner's Sons, March, 1902). I venture to quote it here that it may be compared with the conclusions of the present text.

"The universe of science is, when all is said, a cosmos which is not only true, but also beautiful, and in some sense good. Science tells us what is true; that is science's prerogative: and whatever may be science's final word about nature, that word is in so far the truth of the matter. Philosophy then enters her questions: How can such truth be also good, beautiful, livable? . . . While others say other things, and many others many other things, I say—using the liberty of this preface—It is good and true *because it is beautiful*. Nothing can be [finally] true without being beautiful, and nothing can be in any high sense good without being beautiful. In the words of my colleague and friend, Professor A. T. Ormond[2] (*Foundations of Knowledge*, p. 228), 'the æsthetic principle is at the same time a demand and an intuition . . . an ideal requirement and an intuition under which our world completes itself. . . . It repre-

[1] Citations from these unpublished lectures are made by Dr. (now President) W. D. Furry, in his monograph, *The Æsthetic Experience, its Nature and Function in Epistemology* (Baltimore, 1908), a work in which the author, a member of my philosophical Seminary at Baltimore, carried the pancalistic point of view into the interpretation of certain epochs in the history of philosophy.

[2] This allusion to the views of Professor Ormond is now recalled and made more explicit on p. 215 below, and in Appendix B.

sents the point in our conceptions where worth and truth coalesce and become one.'

"The ascription of beauty, a reasoned, criticised, thought-out ascription of æsthetic quality, is the final form of our thought about nature, man, the world, the all."

This profession of faith is now given its justification after a dozen years.[1]

It has seemed to the writer that this volume might serve as a sort of Introduction to Philosophy, at least to those who conceive of philosophy not as in opposition to science but as inclusive of it. For the developments herein proceed upon conclusions drawn from psychology, theory of evolution, logic, sociology, anthropology, ethics, æsthetics, and history and science of religion; and in each connection the summary given is made intelligible apart from actual resort to the detailed treatises on those subjects. The final conclusion is reached as one resting upon the sciences, but as one finding its place in the historical series of interpretations of nature and man to which we give the name philosophy. The historical sketch given in Part II. and the conclusions arrived at in Part III. show this. The method, also, while new in some of its applications, is

[1] The following explicit statement from the *Psychological Review*, May, 1904, p. 219, summarizes the pancalistic doctrine: "There is here [in the æsthetic] a higher psychical immediacy in which all the dualisms of the mental life, at the stage reached, may on occasion merge in an immediate contemplative value of real presence; the dualisms of 'theoretical and practical,' 'mind and body,' 'inner and outer,' 'freedom and necessity,' all merge to the vanishing point in the æsthetic." The article from which this citation is made gives a table which presents to the eye the main lines of development—the "progressions"—of the mental life which were worked out later in detail in the work *Thought and Things*.

stated and defended in relation to other methods of philosophical procedure. The discussions of æsthetics and of the primitive and unlogical interpretations of the world may be found to bring something fresh into philosophical introductions in comparison with the many restatements of the problems of philosophy in logical form of which our literature is full.

As adding an additional feature of utility to the volume in this direction, an extended Glossary is added presenting the definitions of terms new and old to which the genetic point of view has given a precise connotation.

J. M. B.

September, 1914.

GENETIC THEORY OF REALITY

PART I: INTRODUCTION

GENETIC INTERPRETATION

CHAPTER I

THE PROBLEM: GENETIC MORPHOLOGY

§ *1. The Question of Interpretation*

IN an earlier volume,[1] in which the programme of the entire discipline of Genetic Logic was outlined, the following sentence from Lotze was quoted as defining what we are now to consider as the problem of interpretation.

1. "It is," says Lotze,[2] "the question how far the most complete structure of thought which all the means enable us to rear, can claim to be an adequate account of that which we seem compelled to assume as the object and occasion of our ideas." This pronouncement is fully applicable to the topics which are now to come beforc us. Assuming the completed structure of knowledge as genetically made out, "the complete structure of thought[3] which all our means enable us to

[1] *Thought and Things*, vol. i., "Functional Logic," Introduction, sect. 16. The place of this problem in the whole scheme is spoken of both in the general Introduction as just cited, and also in the Introduction of volume iii. of that work, which presents the topic of Real Logic. In the volume last cited, entitled "Interest and Art," the problem of Genetic Epistemology is treated, while here the correlative problem of Genetic Morphology is taken up. See Appendix A.

[2] Lotze, *Logic*, Eng. trans., vol. i., p. 12.

[3] Rather, in our case, the complete structure of "experience," not merely that of thought, narrowly understood.

rear," the further question then is, how far this complete structure of thought or apprehension—pre-logical, logical, and hyper-logical—can claim to render adequately the reality which "we seem compelled to assume as the object and occasion of our ideas."

This question is sufficiently definite and not at all unfamiliar. It falls apart naturally into two: first, that of the "claim," to use Lotze's term, that thought or experience makes to render reality, the problem of Epistemology[1]; and second, that of the validity of the results—the success or failure, in this task, of the different motives and processes of experience. How are their respective results to be adjusted in the meaning of reality present in the assumptions of knowledge and life alike? This is the problem of Comparative Morphology, which follows upon that of Epistemology, both utilising the results of genetic research. Together they constitute "Real Logic."[2]

2. This last is the problem of interpretation. We have, as Lotze intimates, the natural and ordinary assumptions, in life and experience, of what is superficially called "reality." But we go on to reach, always and everywhere, ways of understanding, construing, thinking about, this reality; in some cases, from mere curiosity; in other cases, for purposes of practical utility. The savage who falls on his face in the presence of the eclipse, no less than the astronomer who turns

[1] Epistemology, taken in the broadest sense, as including all possible modes of apprehension, affective and conative as well as cognitive.

[2] "It remains for Real Logic (1) to interpret each entire series of objective constructions in terms of the sort of reality which their cognition attributes to them; and (2) to *interpret all of them together* in the mode of reality in which their common cognition terminates, *if there be such* a comprehensive mode."—*Thought and Things*, vol. I., chap. i., sect. 11.

his glass upon it, is "interpreting" reality. The action of each shows how he understands the event, what its meaning is to him.

§ 2. *The Historical Problem of Interpretation*

3. Our method, however, requires that the problem be stated in historical and genetic form. We are led to see that there are not only typical sorts of interpretation appearing as the outcome of typical stages and modes of mental process, but also that there are recognisable genetic reasons for their appearance when and where they do. A primitive man can not understand the world in terms of meanings of refined reflection, any more than a child can understand the explanations given in the text-books of physics and psychology. We may suppose, therefore, that these differences have themselves genetic order and development, that there is a natural history of interpretation itself: a continuous movement in the adjustment of the motives of the apprehension of things, resulting in a progression in the evolution of world-views, with the evolution of human culture. If the individual's typical experience passes successively through a series of modes of apprehension—projective, religious, dualistic, logical, æsthetic,—we should expect the history of culture to reveal a similar progress. Similar stages and points of view should be observed in the development of racial interpretation and reflection.

This is a restatement, from a genetic and evolutionary point of view, of the problem of "stages" of thought stated by August Comte, with a suggestion as to its solution. What have been the stages in the evolution of interpretation? And why—from what genetic motives—have they been what they have? If the development of the apprehension of self and things, in the individual's case, is correctly made out, we may be able to trace out the correlative stages in the historical evolution of the race. The social products could not rise higher, at any period, than the individuals' knowledge and practice would justify—so much negatively. And the notable and typical gains of individuals, in mental and moral progress, would show themselves, in time and in serial order, in the texture of the social fabric; in its institutions and in its theoretical speculations—so much, at least, more positively.[1]

4. The consideration of the question of the history of interpretation falls to the genetic theory of reality and constitutes, in our plan, the first great branch of Morphology. Our consideration of it takes form parallel to that of the consideration of the meaning of reality to the individual[2]: the types of interpretation found in

[1] In my *History of Psychology: A Sketch and an Interpretation*, this thought is carried out, the development of psychology being considered as the evolution of the dualism of mind and body and its interpretation, parallel with a similar movement in the individual. That little work (London, Watts; New York, Putnam, 1913) was written and published in the interval between the appearance of vol. iii. of *Thought and Things* and the present volume. It develops intimations made in earlier publications and worked out in university lectures at Baltimore, 1903–6. The same thought has been utilised in treating a special problem in the monograph of Furry, *Aesthetic Experience* (Psychol. Review Philos. Monographs, No. 1, 1908).

[2] The "Genetic Epistemology" of *Thought and Things*, vol. iii. Of the two more evident explanations of the parallelism or "concurrence" between the two movements, this seems the preferable, because it is more empirical and demonstrable. It suggests a *de facto* relation, the presence of influences by which the individual and social apprehensions have actually advanced together and in reciprocal relationship to each other. . . . The other (the "rational") explanation assumes a certain constitution or "nature" of thought, which shows itself to be the same in the two cases.

the history of thought are, in genetic order, pre-logical, logical, and hyper-logical. This intimation may here suffice for the purpose of introduction to the detailed investigation itself (Part II., below).

§3. *The Intrinsic Problem of Interpretation*

5. The problem of interpretation itself remains over, apart from the question of history. The consideration of its history is ancillary to its own treatment. We have, then, as the second and last great problem of Morphology—as it is the last also of Genetic Logic as a whole—that of the interpretation of the meaning of reality as legitimately issuing from the entire mental movement whose development has concerned us. Which, if any, of the historical interpretations is justified in its results?—what does each motive contribute to the full meaning of the real?—what is the synthetic and reconciling mode of apprehension in which the movement of commerce with the real itself overcomes its own dualisms and oppositions, and attains a fully satisfying contemplation of that which, in the words of Lotze, is "the object and occasion of our ideas"?

This is the intrinsic problem. Our treatment of it will be genetic in the sense already illustrated.[1] Our aim is to follow the movement of mental process and not to dictate to it; to observe it, not to direct it. It is in this that the treatment of the topic as a branch of

[1] A recent able discussion of Genetic Morphology, taken in this sense, is that of W. M. Urban in the concluding chapter (xiv.) of his work, *Valuation: Its Nature and Laws*. As is indicated later on (chap. x., sect. 19), this is the fundamental problem of the "critical" enquiry of Immanuel Kant, with the important difference, however, that our method is genetic, not "critical" or logical.

genetic logic differs from the traditional, purely logical and theoretical, discussions. These indeed belong, as we are to see, to a special period—the "logical" period—in the history of interpretation.

But there are other motives entering into the full meaning of the real, beyond the discursive or logical, and it is equally our business to discover them.

6. As has been said, the historical movement of interpretation shows the three great periods conveniently characterised as prelogical, logical, and hyper-logical; and these three headings will be our divisions in the treatment of the historical movement.[1] But in the consideration of the comparative factors entering into the interpretation of the meaning of the real as such, we find that such a division—based on successive cross-sections, as it were, of the entire content of mental process—is not adequate. It indicates the character of the psychical process from which the interpretation issues and the general type of solution as dependent upon the mental function. But it does not penetrate of itself into the solution and discover the motive at work by which the type of reality is determined. At the same time, by its distinction of modes of process—as being prelogical, logical, and so forth—it gives us hints by means of which we may utilise the results reached in the examination of the processes themselves.

Such, indeed, has already been our resource: we have found, in the discussion of Epistemology already referred to, that the meanings of the real persistently fell into two great classes. On the one hand, both the demands of action issuing from the motives of the

[1] Except that instead of "hyper-logical" we speak of "immediacy" theories, a term more significant for classification as explained below.

practical life, and those of knowledge proper issuing from the motives of the intellectual life, proceed, always and everywhere, by a process of *mediation by and through representative states of mind or ideas.* On the other hand, the interests of immediacy, of direct apprehension and contemplation, are present in characteristic forms in the lowest and the highest reaches of conscious process, in the primitive and in the hyperlogical. All reality, therefore, whatever the process of reaching it may be, is either immediate or mediate, either immediately given or mediated by ideas. If this be true, our problem is greatly simplified. We have to ask, not whether this or that stage of development gives the final interpretation of reality; but which type of "real" meaning, the mediate or the immediate, is the more comprehensive and satisfying.

In this part of our work therefore (Part III) we will examine interpretations based respectively on Mediation and Immediacy, taking account of the several types of each revealed in the historical account of the stages of interpretation (Part II) and reaching our own reasoned results.

In the last Part (IV) it will remain to gather up certain Conclusions and Corollaries of a philosophical sort, which illustrate the wider bearings of Pancalism.[1]

§4. *Scheme of Treatment*

In view of the foregoing, we arrive at the following scheme of treatment:

[1] Certain Appendices follow, of which the first (Appendix A) may be especially referred to here, as well as a "Glossary" of the terms used in this and other genetic treatises. On the term "Pancalism," see the remarks in the Preface.

TABLE I

Genetic Morphology: the Problem of Interpretation.

Part I. Introduction.

Part II. Historical: the Development of Interpretation.

i. Early Racial (pre-logical and religious) Interpretation.

ii. Logical Interpretation: theories based on Mediation.

iii. Hyper-logical Interpretation: theories based on the Immediate.

Part III. The theory of Æsthetic Immediacy: Pancalism.

Part IV. Conclusions and Corollaries.

CHAPTER II

INDIVIDUAL INTERPRETATION

§ *1.* *The Nature of Interpretation*

1. It will become plain as we proceed that what we are calling Interpretation—following the usage already suggested[1]—is generally covered by the term "meaning." The best way to introduce the topic, indeed, is to recall briefly the theory of meaning as it has been worked out in recent discussions, and take advantage of what we know about it.

In general, interpretation is simply the entire meaning given to an experience, fact, event, by the consciousness which, as we say, "makes" the interpretation. Given bare happenings—a flash of light, a rumbling sound, a shaking movement—each item *means* this or that experience: a thunder-storm, a passing cart, an unsteady chair. Whatever intent of added meaning attaches to the bare content of fact or idea, it is in so far a personal interpretation of the fact or idea. It may be little or much, superficial or profound. Infor-

[1] This chapter sets together, under the heading of Interpretation, the results of earlier detailed discussions, so far as they are comparable with those of racial thought as brought out below (Part II.). It aids us also in the task of focussing our thought in the final interpretation to which we ourselves arrive (Parts III and IV). See also *Thought and Things*, vol. i., and Höffding's *La Pensée humaine.*

mation, purpose, utility, association, habit, appetite, instinct—a dozen motives of meaning—may surge up and segregate themselves about the nucleus of the given datum. All the experiences cited as examples above, sensations of sight, hearing, muscular sense, may be united together in the interpretation made by the Sicilian, who cries "earthquake," and rushes from his dwelling in terror. And a similar interpretation may be arrived at by the scientific man seated at Washington, who has before him simply the tracing of a needle on carbon paper, hearing, seeing, and feeling nothing of the physical event itself. His interpretation is more remote, more indirect and logical, as we say, but it expresses, just as the Italian peasant's does, the meaning he attaches to the datum that comes before his mind.

2. Avoiding debated questions, of the nature of the datum, the relation of content and intent, of presentation and interest—all questions elsewhere sufficiently discussed[1]—we may here simply point out that the interpretation, considered as attaching to the thing interpreted, is of the nature of "intent," over against content. It has the variability and personal quality—as being relatively mystical or logical or æsthetic—given it by the individual's mode of apprehension and feeling. As meaning, it is selective. Over against this, there is the relative constancy and recognitive character of the content, the common and neutral thing or event that is interpreted. The sound, the light, the movements of the earth, are heard, seen, felt, in about the same way by all normal persons; they constitute the same "content." But the interpretations of one or more of these events differ seriously. The actual event may give rise to all sorts of conjectures and

[1] See "Functional Logic," vol. i. of *Thought and Things*.

suppositions, each in its way a tentative "intent" of interpretation.

3. In the tangled wilderness of these differences, however, it will pay us to attempt to discover pathways —lines of classification and distinction in the motives and in the results. And it appears that we have at hand means of doing so. The general distinction between two great factors of knowledge, the datum and the interest, should be recalled; and the question asked as to the relative place and rôle of each of these in this or that case of interpretation. On the one hand, the datum goes on to develop into the objective reality which controls the manifestations of interest and action; on the other hand, the interest gives personal and selective meaning to the objective system of data. Let us consider the nature of interpretation somewhat more fully from this point of view.

§ 2. *Interpretation as an Organisation of Interests*

4. The interest which the individual brings to any objective experience is as varied as are the motives of action and knowledge. As has often been pointed out, the simplest "thing"—the sapling, say, at the roadside—may be rendered in a variety of ways, which testify to the dominant habit or to the casual interest of the observer. The modest tree means the refuge or shelter, the source of fire-wood for the home, the specimen of horticulture, the value for exploitation, the species of botanical science, the object of æsthetic pleasure, the symbol of religious truth. All these are legitimate and synchronous interpretations of what is, in the experience itself, a certain controlling set of conditions—controlling upon the senses, upon direct

observation, and upon social acceptance. We say to one another, "See that tree!"—and with the reply comes the possibility of reporting and confirming in common the basis upon which the different interpretations all depend. These interpretations, however, may run away with the datum, so to speak, piling upon it a mass of selective and personal interest, which obscures and distorts it. Indeed the great motive and the prolonged struggle of the cognitive interest, as such, the interest in neutral and objective truth, is just to disentangle the bare object from the meshes of preferential interest and interpretation, and preserve intact its system of commonly observable relationships.

While we may say that from the first the individual's principal types of interest are germinating and are already more or less organised in the ready mechanism of appetite and instinct, still it is evident that there is a genetic progression in the life of interest. In the stretch of mental development, up to the dawn of the explicitly logical, it is not the cognitive interest that is predominant and directive. On the contrary, the exigencies of practical life—indeed those of physical existence itself—are controlling upon the apprehension and interpretation of things in the world. All things that serve to pick up milk are "spoons" to the child, and any one who sings to him in the dark is "mama." The lines of cleavage are not drawn by the hand of sensation, but by the hand of interest, practical and emotional.

5. The emotional, as such, takes on great importance with the development of the child's natural affections. Attachments of a quasi-instinctive sort dominate situations in which many alternatives are otherwise possible: the great social motives of family, clan, and race. We have examined in detail, in our discussion of

"affective logic,"[1] the principles under the rule of which the great emotional preferences and interests are organised in relative independence of the organisation of the motives of knowledge, though not in actual separation from them. The distinctively practical or utilitarian motives, subconsciously at work, are supplemented by those of emotion and affection, both tending in their outcome to silence or render inarticulate the voice of cognition as such.

6. The more special and seemingly exclusive organisations of interest should not blind us, however, to the presence of other and less exclusive ones, proceeding along with them and perhaps using the same objective data. The child who does not seem to distinguish the real spoon from the hollow building block, for the purpose of eating, does not think of using the spoon as a building block. He perceives, then, clearly enough, the actual difference of form. In spite of the varying larger interests which in turn obscure the simple shape or form, this latter is nevertheless on occasion properly isolated. The boy who does not, because he need not, distinguish between his mother's and father's voices when he is called to dinner, does make the distinction very promptly when one or the other summons him to get up in the morning! So, as we will have occasion to see later on, the savage who seems to have no adequate perception of the mere club or feather as such, merging the sense-data in a larger mystic interpretation of their meaning, still uses the weapon or the ornament on occasion under another interest, in a way which reveals an accurate apprehension of its form, size, colour, and physical properties.[2]

[1] In Part III. of "Interest and Art" (vol. iii. of *Thought and Things*).

[2] This is a very significant fact in connection with the interpretation of savage modes of thought; see below, chap. 4, § 3, sect. 19.

There is, then, in the working of the mind as a whole, even in the early stages of apprehension, a great variety of possible organisations of interest and data. In one case one dominant interest comes to the fore, in another, another. The proper and accurate cognition of things of sense may coexist with the organisation of interest led by instinct or by association of sense with sense; and with both of these there may exist larger practical and emotional complexes. On occasion, as an independent issue or as means to a further end, the datum, the bed-rock of perception, which, as a body of objectively stable and unyielding objects, is the control upon action, comes sharply into its own. Generally, however, it serves only as a condition, a sign, a trigger, a signal for the outpouring of the mass of affective processes and of social habits, which overflow and conceal it. It is only in the stage of logical process, dominated by judgment, that the cognitive motive and interest, as such, become in any sense independent and autonomous.

7. The important thing to note in this connection is that the mental life is often given over to types of interpretation in which affective and active factors play the predominant rôle. The factor of cognitive apprehension, as such, is present in interpretations of this type, but it does not, in the prelogical stages of development, assume independent form and become itself a predominant interest. The term "prelogical" means pre-discursive, pre-theoretical, pre-scientific; it does not mean pre-noetic, nor in principle anti-logical. The motives of knowledge are present, but it is not mainly in their interest that the mental life is organised and objective data interpreted. It is the subjective factor, not the objective factor; the interest, not the datum;

the value, not the truth, that actuates the movement of the mind.

This receives further on a remarkable emphasis in our consideration of the social interpretation to which the data of early knowledge submit.

§ 3. *Interpretation as the social Organisation of Interests*

8. The inestimable debt that individual knowledge owes to society has recently been pointed out. In our detailed examination of the "common" force of knowledge,[1] at its different stages, we have come upon abundant evidence that it is by social or "secondary" conversion processes that individual results are confirmed and corrected. Thus the vagaries of personal preference and private idiosyncrasy are discovered and discounted. The individual submits to the common and social verdict, and by it learns to think true.

It has not been made plain, however, that the opposite may be sometimes the outcome: the individual may be the victim of social errors, of current vagaries. The assumption that the individual's results are corrected and made reliable by social tests rests upon the further assumption that the social results by which they are tested are themselves correct and reliable. But what guarantee has society that its results are reliable but the resort in turn to the experience and judgment of individuals?[2] The best that society can do for the individual is to bring him into agreement

[1] See *Thought and Things*, vol. i., chap. vii. (in general) vol. ii., chap. iii. (common force of judgment); vol. iii., chap. vii., (common force of practical rules).

[2] The writer has dwelt upon this revision of society's truths by individual tests in the work *Development and Evolution*, chap. xvii. See the diagram given in *Thought and Things*, vol. ii., chap. vi., sect. 27.

with itself; but the result may be right and it may be wrong.

9. The individual becomes aware of this bondage to the social when he learns to judge for himself and to have resort to personal and direct experience of nature and man—when, that is, he resorts to "primary" instead of to "secondary" conversion tests. Valuable as the social training is in matters of practical life, and in all knowledge that represents accumulated experience and is ready to submit itself, or has done so, to the immediate tests of trial and criticism; beyond this point it becomes a "blind leader of the blind." As matter of fact, the individual spends much of his period of competent reflection and private judgment in correcting the errors of belief and practice into which his submissive social training has led him.[1]

The result is that the entire body of socially accepted and socially enforced "truths" are subject to gradual revision and correction by the thought, invention, and discovery of individuals. The more inventive thinkers insist upon bringing to the test the established and conventional formulations in which they have been reared. The social view is found to have arisen by pro-

[1] As knowledge becomes mature it passes through the period of enforced commonness, the result of which is merely "aggregate" (in the terms of the discussions referred to), on to the period at which the value of agreement and common conformity becomes apparent to the individual and he consciously imitates the current models. His thought is then "syndoxic." Both these periods represent the social apprenticeship of thought. But when judgment arises, the tables are turned. The individual serves notice upon other thinkers to agree with him. His judgment recognises its kinship to the judgment of the social fellows, as reaching the same conclusions. Let us all cease imitating one another, says he, cease accepting anyone's "say-so"; let us resort in common to the source of all sure information, the facts as they are. Judgment, in this sense, is "synnomic." So also are the rules of practice issuing from moral judgment, and for analogous reasons.

cesses of prelogical apprehension, mystic participation, and imitative propagation rather than by the tried and tested methods of actual experience and experiment.

Thus society is carried out of the pre-logical into the logical stage of thought, through the personal competence and insistence of individuals. Society is the disciplinary agent, the schoolmaster, to the individual's thought; but the pupil outgrows the social school. He learns that he and the society alike, of which he is a part, have to submit to the tests of another and impersonal system of controls: those of nature, given in perception; and those of logical consistency, given in the body of reasonable thinking.

10. In the prelogical stages of knowledge, interpretation shows the lack of logical coherence and organisation. The immaturity of thought shows itself in respect to each of the criteria of the logical mode as such. The self of clear self-consciousness being undeveloped, there is the blending of the individual's personal interests and actions with those of the social group; a mutual interpenetration of selves, as social habit and common interest require. The self of family, clan, tribe, representing a solidarity of interests within the circle, forbids the assertion of individual desire, will, or caprice. All is ruled in the interests of the group. It is only gradually that the youth, even in civilised society, learns the lessons of individualism and sturdy self-assertion.

In the matter of the interpretation of external things also, the immaturity of judgment shows itself, through the acceptance of the social decree in place of the testing by fact. As long as the dualism of idea and thing, of the subjective and objective, is not clear, the interpretation of the external may allow all sorts of interpenetration of things and selves. Things may act as selves do, the caprice of persons may attach to things, the mystery of the explosive and uncaused actions of persons may extend to all the world. This leads to a mystical and emotional interpretation—"fearful," in the religious sense—of the commonest objects of the external world.[1]

11. With these things go the characters of pre-logical knowledge which put in evidence the absence of the faculty of judgment. It is by judgment, together with its rules of consistency, contradiction, sufficient reason, and so forth, that the system of logical truth is built up. In the pre-logical modes these rules and restraints are not present;[2] the motives of suggestion, imitation, emotional contagion, personal imagination, social impulsion, religious fear, work out their proper results. Illogical and unlogical things pass muster; the child, like the savage, accepts the wildest tales of fancy, the most impossible violations of time, place, and cause, without flinching in his acceptance or revolting in his emulation. The stories told children of religious miracles and feats of national prowess, pass through their minds with the simple *visé* of teacher, story-book, or well-informed friend. Why not? The children are organising their interests, making their interpretations,

[1] We have seen in some detail (in *Thought and Things*, vol. i., chap. v., sect. 12, 24) that the confusion is exemplified in the case of the apprehension of his own body by the individual person. The body is a centre of mystery. It embodies both sorts of reality at once: it is both a person and a thing, the locus of physical change and the fulcrum of personal effort. If this physical body can be the seat of pain and effort, and the theatre of all the personal dramas of dreams and reveries, why may not other bodies also? The personal body is the centre of many of the most remarkable rites and mysteries of primitive belief and usage.

[2] We have seen, in our detailed investigation of emotional organisation, that affective logic lacks the principles of "contradiction" and "excluded middle." See *Thought and Things*, vol. iii., "Interest and Art," chap. vii.

as best they can; their properly cognitive and logical apprehensions are not controlled apart from their affective and active impulses and dispositions. The latter, the un-logical, are more urgent, through the constraint of the social life, which puts no premium upon individual thought—just the contrary. The result is that the interpretations of the individual, like those of the group, are dominated by the emotional and uncritical acceptance to which the term "mysticism" is generally applied.

§ 4. *The Progression of Interpretation*

12. Thus understood, interpretation is nothing more nor less than the process of reaching objective meaning. It passes through the genetic periods designated prelogical, logical, and hyper-logical. In the detailed treatment of these stages, movements of a semi-logical or quasi-logical sort have been found, bridging the gap between the prelogical and the logical; and also, similarly, processes of a vanishing logical or over-discursive sort, leading from the explicitly logical to the higher intuitive and contemplative. The entire series, for our present purposes, is as follows: perceptive, intuitive, semi-discursive, discursive (logical), over-discursive, contemplative (intuitive).

Considering this series, in a broad way and from the point of view of function, we note certain interesting things. The apprehension moves from one intuitive pole, the intuition of direct apprehension, to another, the intuition of contemplation. Between these lies the vast tract of discursive or logical process, in which the directness of intuition gives place to the mediating processes of thought. The acceptance of sense-perception and feeling yields to the mediated belief of judgment and reflection; and these in their turn find their supplement and completion in the acceptance of rational intuition and æsthetic contemplation. Through thought there is a freeing of the mind from the inadequacies and inaccuracies of first-hand and uncritical acceptance; processes of personal judgment and logical grounding succeed the ready formulas and conventions of reality-feeling and social convention. But this is not all; there is a second freeing. Motives of an intimate nature—the "private" values, ideals, and ends—together with the "singular" values of emotion and interest, survive and claim more adequate place than that given them in the indirect and discursive reports of thought. Accordingly, the *élan*, the effort of the mental movement, is toward a higher immediacy, in which the gains reached by discursive process are integrated in the new directness and immediacy of contemplation. This is the second freeing of the mind: the freeing of the faculties of knowledge and value alike from the trammels of thought, from the machinery of mediation.

These two "freeings" occur at the transition periods spoken of above. It is worth while to consider them more closely in view of their importance both in the individual and in the corresponding racial movement of interpretation. They also figure in the final conclusions to which we come in the last part of our work.

13. The first of these transitions, called just above that of a "freeing through thought," has the semi-logical characters which show the essential motives to all mediation, cognitive and active alike. It comes by a flowering, an expansion, of the imagination. Images are utilised as the medium of truths and ends; they are set up, played with, schematised, manipulated.

The imagination projects, postulates, conjectures, makes experiments, all by means of images. "As-if" constructions, semblant in character, have their turn; make-believe runs ahead of belief; playful treatment of results poses as experimental confirmation; fairy-tales serve as introduction to the wonderland of reality. The individual has his fancies, and the race has its myths—both handmaids to thought.

In two ways does this flowering of the imagination serve to form the motives of thought. It acts by an assumption of reality, which foreruns the grounding of fact, and is instrumental to belief. It is the midway point between the immediacy of direct presence found in sensation and perception, and the full immediacy of the conclusion confirmed by actual testing and by the reducing processes of logic. And it is also freeing in the sense that it clarifies the inner realm in which images of all kinds disport themselves. It thus prepares the way for the explicit dualism between the self and the objects of its thought. Both of these results have been brought out in detail in our discussions of individual development; they are both plainly to be found operative also in the racial movement of interpretation, as will be shown below.

14. The second of the periods of transition, spoken of above as being also a "freeing," is worthy of attention. It shows the transition from the logical or discursive, as such, to the hyper-logical, higher intuitive, and contemplative. It is a freeing *from* thought, as the other is a freeing *of* thought. The use of cognitive schemes and logical concepts, of ideas and conceptual categories, belongs to the method of thought; it is its resource to such an extent that consciousness becomes accustomed to exercise its acceptances and rejections

on grounds of reasonableness alone. The result is that a disparity arises between the world of immediate experience, given in feeling and action, intuition and presentation, on the one hand, and that of mediate or grounded results, on the other hand, whether they be in the realm of truth or value. Thought is limited in its rendering of experience; the singular and the immediate escape its net. The reintegration of these with the results of thought in the whole of experience become then the task—and to us the problem—of interpretation.

15. This task is accomplished, in the natural movement of consciousness, by another and very remarkable exercise of the imagination. It is by the imagination, as has been said above, that the mediation of thought is first ushered in. So it is by the imagination, also, that the limitations and restrictions of thought are overcome. In the former case, the forecasts of the imagination pass into the confirmed and controlled results of mediation. In the latter, on the contrary, the imaginative scheme, the semblant forecast, maintains itself, as over against the claims of both types of mediation. The self refuses to retire into the sphere of subjectivity, over against the objective and external. On the contrary, it reads its presence into the experience which is already enriched by the accretions of mediate process, claiming to secure, by direct contemplation or intuition, the full realisation of the subjective meaning.

In the individual, this freeing of experience from the restrictions and limitations of thought, brings a new equilibrium to the whole mental life. It negates the exclusive claims of rational and voluntaristic process alike, and restores to the immediacy of feeling its dignity and value.[1]

[1] Upon these hints, contained in the spontaneous process itself, the interpretation of reality given below, in Parts III. and IV., is based.

We find below, in the sketch of the history of philosophical interpretation, that the extreme logical or rationalistic solutions of the problem of reality have been accompanied or followed by movements toward the recognition of immediacy, in various degrees mystical and æsthetic in its character.[1] As the earliest dualistic forms of theory were ushered in by departures from the immediate acceptance of things as they are, criticism succeeding realism and reason coming after superstition; so in turn these latter reveal motives of more refined dualism, the terms of which can be reconciled only in a new synthetic and æsthetic unity.

Here, however, our purpose is to recall the fact that in individual experience this is the case: in the æsthetic experience, the motive of essential reconciliation is to be found. It is not achieved by the undoing of the work of mediation; but, on the contrary, by the integration of its results in the larger and richer unity of immediate contemplation.

§5. *The Realities as Interpreted*

16. We have now thrown into relief the outline of the movement of the interest which is active in interpretation[2]; it remains to recall briefly the stages reached in the development of the other great factor, the objective datum.

First of all, it is to be recalled that experience does not at the start isolate and distinguish the two sorts of reality represented by the dualism of subject and object, of minds and things. On the contrary, experi-

[1] This has been shown in detail by Furry, *Æsthetic Experience: its Nature and Function in Epistemology*, Part II.

[2] The details are given in the work mentioned, *Thought and Things*.

ence is merely "projective"—a projection before the gaze of the mind of a confused mass of striking happenings. Gradually out of this all the dualisms of later meaning emerge. The reality of a presentation at this stage is simply an aspect of the presentation itself. It has been called "reality-feeling," in contrast with the belief proper to the later period, when judgments of truth and existence are possible. In the terms of recent discussions of the subject, reality is simply "presumed"; it is not judged nor is it even assumed; for there is no thought of reality as in any sense attached to or separate from the presentation in question. The presence of the content is all its reality means; its apprehension carries with it the presumption of those elements of persistence and stability which go on to develop into positive control. Existence and reality are later meanings, developed from this primitive presumption.

Such is the reality of the datum in the prelogical period of knowledge. It is a meaning of immediate presence and intuition.

17. In the ensuing period, the motives of dualism and discursive process lead rapidly on to the logical mode; and a similar transition takes place in the meaning of reality, considered as datum of existence and control. The imagination takes the lead, as we have seen, projecting its schematic and tentative readings forward with various shadings of semblance and probability. To it, of course, the meaning of being real is not a mere presumption, since its force varies for different cases, and since it works out varying results under the tests of fact and truth. But before this testing, it is not the settled conviction of grounded belief; it has the character of assumption, proposal, hypothesis. The interpretations put forward by the

imagination are assumed to hold, in the pursuit of the interest by which the process is motived. Reality is embodied in all sorts of "as-if" constructions, from the make-believe of play, together with the pretence of social *mensonge*, to the sober hesitation of doubt and the capricious will-to-believe of passion. The entire period in which the dualisms of self and other, of mind and body, of substantial existences, of subject and object—the period of the freeing of the motive of judgment from the bonds of social convention and authority—is shot through with the varied assumptions of possible existence and activity. All its interpretations have the force of possibility and probability, which discounts in advance the realities of judgment and proof.

18. The logical type of reality appears when experimentation in its many forms serves to establish or to annul the true or false assumptions of the imagination; when the period of logical process proper—judgment, reasoning, implication—is ushered in. The germinal distinctions of control are hardened into the categories of existence and reality. The two substances, mind and body, are set up over against each other; the self retires into the citadel of subjectivity, distinct from all its thoughts; the ego-self becomes, both in thought and in action, the reflective and self-assertive individual. The judgment, with its synnomic force, holding all to the results of each, succeeds to the uncritical presumption of social custom and habit, and to the schematic assumption of the imagination. As on the functional side, there is a releasing of the faculty of thought, so on the side of content or reality, there is a crystallising of the meaning of existence in the categories into which thought casts its objects. Thought, logical process,

in short, comes in to settle the questions of possibility and probability of the semblant imagination and to decide what is to be accepted as real.

Thought settles these questions by means of the two great processes of mediation to which it is always committed. By the logical mediation of ideas, it reaches conclusions that are true; and by the teleological mediation of means, it reaches ends that are valuable. The question "what?" is answered by the one, the question "what for?" by the other.

In this mode, therefore, all objects or things have definite predicates of existence and value; they are what they can be proved to be, in both respects. They are also involved in a system of implications—causal, temporal, spacial, and so forth—which serves the unifying and relating interests of thought. The reality of each sort of existence becomes the presupposition of the object's acceptance; belief with the presuppositions of reality succeeds to the acceptance with presumption and also to the semblance with assumption of the earlier modes.

19. The logical, however, does not appear to be the last word. Consciousness achieves a freeing from logic as before she worked to secure the freeing of logic. The dualism of substances, that of self and thing, that of subject and object, all bring with them various embarrassments. The categories of thought, by their very nature, ignore and disparage certain of the most direct and pungent experiences of reality and of life. The ideal of the process of thought is one of mediation, as that of volition is; both are ideals of certitude and actuality: but just in those singular and concrete experiences in which directness and certitude are at their highest, the general categories of thought and the

universal readings of logic are formal and inadequate. In them reality is not exhausted.

20. The movement of apprehension then sweeps over and beyond cognitive process and reaches a new immediacy. It embodies itself in a new *aperçu* of reality. The process of discursive thought, on the one hand, erects its principles into intuitive truths, absolute data of reason, from which the indirectness, remoteness, and discursive quality of thinking have disappeared. The presupposition of logical implication becomes the postulate of pure reason; and reality discloses to rational intuition its intimate nature. Thus the intuition of the real takes on rational form.

On the other hand, the mediation of the active life pursues a similar ideal; it, too, seeks to secure absolute and final validity for its injunctions of moral conformity. Its ideal is attained also by a new immediacy and directness, an intuition of the rules of practice; but it is attained by a procedure the reverse—as we have shown in detail[1]—of that of the theoretical reason. It erects into postulates its ends, values, and goods; and drops out the instrumental means by which these goods are secured. Conditional and relative ends become absolute values, seeming to be imposed by the very nature of things, and to disclose the essence of the real as a system of practical goods.

21. In this twofold way, then, the interpretation of the real takes on a new phase: it passes from the sphere of discursive thought into that of over-discursive intuitive apprehension. But in each of these two ways of doing so, the rational and the practical, only a partial ideal is realised, since they are in strong contrast to each

[1] See *Thought and Things*, vol. iii., "Interest and Art," chap. viii., sect. 18 ff.

other. In one, the ideal is one of rational implication; in the other, one of practical conformity. Each is in so far absolute for its own process and content, but inapplicable to the process and content of the other. The result is, as we are to see fully later on, a deep-seated and unconquerable opposition in the development of the philosophical systems into which interpretation takes on reflective form. Just as, in the individual, mere reasonable appearance does not suggest nor enforce the practical imperative; so in theory the ideal of rational system, embodied in a given set of theoretical categories, does not suggest nor impose the ideal good of the life of desire and will. Rationalism, based on the intuition of "pure reason," finds its rival in Voluntarism, based on that of "practical reason." Interpretation taking dual form in reason and conscience, does not resolve the dualism of truth and value even at the stage of intuition.

22. It remains for the imagination to come to the rescue again here, as at the earlier stage when cognition was choked by its own mechanism. It secures a restatement of both the ideals of mediate process, merging them as moments in a larger mode of immediacy—that of æsthetic contemplation. As on the functional side, the process, freed from the narrow compulsion of argument, returns to an immediacy of direct apprehension, so the interpretation takes up the data of knowledge, together with those of value, in a larger whole of direct realisation.

The meaning of the real as revealed by æsthetic insight is to be fully enquired into below. Here it is only in place to signalise its comprehensiveness, as issuing in a reality of reconciliation and completeness. The real of contemplation does not lack any of the data upon which the ideals of truth or value are erected; it lacks neither the generality of cognition, nor the singularity of sentiment, neither the objectivity nor the intimacy of the poles of self-feeling: all these are included in it. The interpretation which issues from it, therefore, in the history of thought should be genuinely synthetic, not merely eclectic—although it is often counterfeited by eclecticism at the higher stage, as it is by mysticism at the lower. It should be borne in mind that the progression to the æsthetic is a movement of real change in the motivation of knowledge; and that only by showing a similarly genuine necessity of synthesis can the corresponding interpretation be justified on the part of reflection. The movement of consciousness is one of higher semblance, in which the motives of control are brought to a new equilibrium; it is in the lesson of this that the æsthetic interpretation must find its force. Our own interpretation, given in Part IV. below, is based upon this truth.

CHAPTER III

THE PARALLELISM BETWEEN INDIVIDUAL AND RACIAL INTERPRETATION

§ *1. The question of Racial Interpretation*

1. The intimate bond existing between the individual and the group at every stage of the development of the former has been indicated in many recent discussions. This bond is so close, the mutual influence so profound, that the consideration of the individual's interpretation of the world would appear incomplete and partial without that of the corresponding movement in organised social life. The question, how does the individual understand reality?—what interpretation does he put upon it at the several stages of his personal development?—suggests the further question, how does the human group as a whole, the race at the several epochs of its evolution, look upon reality and interpret it? Are there any evidences of an advance from one point of view to another, of historical progress in the social organisation of interest and knowledge? And if so, what relation do these two genetic movements hold to each other? Has there been any sort of coincidence or "concurrence" between the two in the direction of progress and in the order of their stages or modes?[1]

[1] The necessity of including this problem in the programme of genetic logic appears when we recall the results of recent anthropological research as well as the outcome of certain psychological analyses. Re-

In order to approach this question with profit, certain explanations have to be made in advance. We have already come to a clear understanding as to what the individual movement is and comprises; it is necessary to be clear also as to what we mean by the racial or social movement with which the former is to be compared.[1]

---

search in ethnology is showing more and more clearly that in primitive culture the social, the collective, is the leading and controlling factor in thought as in conduct; the individual is a social outcome and his personal competence and integrity are slowly established in the course of social evolution. This is in accord with our results, which go to show the inherent common element in individual knowledge at every stage—the reflection, in the most private judgment, of the social conditions of its origin.

The theory of the way the individual understands the world and his interpretation of it, cannot dispense with that of social evolution, any more than that interpretation itself can dispense with its social support. Every question of individual genesis to-day raises questions of social evolution.

[1] In approaching this question, we should carefully distinguish it from certain other lines of enquiry, with which it may easily be confused. In the first place we are not to discuss the question of the endowment or capacity of races or groups, either in itself or comparatively. Whether mankind has made great progress, little progress, or no progress at all, *in its hereditary traits*, from generation to generation, does not directly concern us. Whether or not the child of the civilised group is better endowed, more capable, than that of the primitive group—the former outstripping the latter if placed in the same social conditions—is not for us to ask. Such an enquiry would follow rather than precede our proper enquiry; for the presence of a given sort of interpretation achieved by this group or that, might, under certain conditions, throw light upon its mental endowment. But even this is not our business. We are to enquire into the actual interpretations historically achieved by mankind, and into the order of their achievement; and then to compare the results, in both these respects, with the achievements of the normal individual in the course of his life history as a conscious person.

Resuming these points more positively, we may say of the problem of racial interpretation that it is a question not of endowment, but of product which, while found in the individual, is not originated entirely by him.

2. At the outset, we may assume that, in that great mass of beliefs we call interpretation, there are not two movements, one individual and the other social; there is but one—the movement in the internal organisation of thoughts and values in the given group. Objectively it is constituted in institutions, in laws, rites, customs, sanctions. Subjectively, it is reproduced in the individuals, who in turn give to the group its vital impulse and its progressive "uplift." In it all the quite fundamental relation between one factor and the other constantly appears.

The concurrent development of individual and racial interpretation is due, therefore, to a single factor, the new achievements of individuals so far as they are generalised and incorporated in the body of social beliefs and customs.[1] The successive stages of social attainment reflect, in their order as well as in their nature, the advances made by individuals. Group thought, for example, passes from a stage corresponding to the prelogical period of cognition into a stage of competent judgment and logical reasoning; but this is possible because individuals have already taken the necessary transition steps and have brought about modifications in the body of social custom and habit. Each such advance on the part of individuals, sufficiently general and compelling to get the social hearing, would produce a progressive step in the interpretation called social or racial; and being then subject to social trans-

[1] Instead of saying as the biologists do of vital recapitulation—that the individual development repeats the genetic series established by racial evolution—we should say that the social evolution has been determined in its several stages by the essential advances made in individual minds. Of course, for continuous individual development the stable and continuous social life is necessary in which the earlier individual attainments have permanently taken form.

mission and concretion, it would remain a sort of stratified racial possession. New generations coming after and building farther—as variations in congenital equipment or individual force and feeling permitted—would repeat the process, depositing anew the increments of further social advance.

These truths would seem to afford sufficient justification of the expectation that a real concurrence in direction, in motive, and in stages would show itself between the individual's mental progress and the progress of the race.[1]

There is, however, a more general way of considering the matter, which may be cited in order to bring the consideration of racial interpretation into direct line with that of the logic of cognition. It is the point of view from which we may speak of it as the organisation of social interest.

§ 2. *Racial Interpretation as organised social Interest*

3. The conception of interest has been found a fruitful one in theories of meaning and cognition; especially so in the consideration of the affective and active motives found in the full apprehension and appreciation of things. We have ourselves[2] found that these motives take on more extended or more contracted forms of organisation about the data of sense; and that in the individual case it is just this organisation which constitutes the "interpretation" of what is given. By the gourmand, the oyster is interpreted as food; by

[1] The writer's little work, *History of Psychology* (1913), takes up the topic in detail (vol. i., chap. 1, and vol. ii., chap. 8), interpreting the history of psychology in the light of the individual's development in dualism.

[2] *Thought and Things*, vol. i., chap. 3.

the scientist as mollusc; by the pearl-diver as desirable treasure: each interprets the given presentation as terminus or satisfaction of his dominant interest.

But we have seen that in all the organisation of the individual's interest, an influence is upon him, amounting to a sort of constraint from which he can not escape. Not only is his perception of the given data of sense subject to obscure influences of social suggestion and habit, but the body of meaning given to the data, indicating the direction of his interest, is shaped in social moulds. The cognitive part of the apprehension is generalised and converted into socially common objects: and the affective parts are read ejectively and imitatively in terms of the opinions and habits of others in a way that secures to interest itself a certain collective force. In the logical field, knowledge has a force of community; judgment is synnomic: and the equivalent result is reached for the intent of appreciation by the less definite but equally real processes of affective logic.

We see here, therefore, the concurrence of individual and social interpretation in its fundamental *modus operandi*. The individual interest is socially construed and confirmed; the social interest is renovated and renewed in the vital interests of individuals. There is just the one organisation—the common intent to find the thing to have this or that meaning. It is an intent common to the group and to the individuals which make it up: to the group, as it is embodied in social tradition and institutions; to the individuals, as it is embodied in their personal beliefs and habits of action. It is constantly re-created—as it was originally produced—by the co-operation of the individual and the group.

4. With this understanding of the nature and origin of racial interpretation, a word may be in place as to the proper rules of procedure under which we should investigate it. The danger, apparent indeed in much of the literature of primitive life and thought, is two-fold. On the one hand, we are likely to obscure the actual differences between stages of interpretation, as of culture in general, by reading into the simpler forms the processes of high individual mentality. The fallacy of the "implicit," pointed out in our discussion of the canons of genetic logic,[1] appears here with pronounced effect. The processes of reflective morality and religion are read into the primitive man's way of understanding the world and of acting upon it.

On the other hand, one is equally in need of prudence and self-restraint in avoiding the other extreme. One who is impressed with the seemingly irrational and mystical modes of primitive thought is tempted to deny the unity of mind in evolution and the continuity of racial progress, finding differences in kind between primitive and civilised man.

5. From both of these errors our conclusion as to the concurrence between individual and racial development should save us. In whatever direction we may find the more primitive types of thought to tend—however partial, illogical, and crude it may show itself to be—we have still to recognise that it has arisen by the organisation of individual interests, in which certain great motives proper to mankind are predominant. The logical interest, as we have seen, develops late. And the prelogical organisation of interest is not ruled by logical principles; it allows wide latitude to affective, mystical,

[1] *Thought and Things*, vol. i., "Functional Logic," chap. 1, sect. 8.

and ejective motives which are reflected in a collective tradition and a social sanction seeming to us to be irrational and superstitious. This being the case, we can always say that however widely given cultures, with their respective beliefs and customs, vary one from another, still each of them shows some predominant human motive working itself out in a characteristic social product. In the shifting and rearrangement of such motives as culture advances, the products are so different that only the larger view enables us to include them all in the one continuous historical movement. The ever-present and sufficient testing and guiding principle, however, is this: each characteristic phase of culture, wherever and whenever found, can have arisen only by the social organisation of essentially human interests in which now this motive and now that has been predominant. The problem in each case is to discover that balance of motives which will account for the facts observed.

§ 3. *The Stages of Interpretation*

6. Assuming this general concurrence—now so far justified—we may exhibit the principal stages of interpretation by means of the following table. In the first part (I) the genetic "stages" of development of the individual's cognition are presented; in the second (II), those of society or the race. The first needs no further description: its justification is to be found in the detailed expositions of the work already cited. The latter constitutes the program of the discussions of the second Part of this volume. In both cases, the transition periods are indicated by the headings at the points of the diagonal lines in the columns marked "modes":

TABLE II

I. Individual Interpretation

Stages	*Modes*
i. Prelogical	i. Intuitive
	Quasi-discursive
ii. Logical	ii. Discursive
	Over-discursive
iii. Hyper-logical	iii. Contemplative

II. Racial Interpretation

Stages	*Modes*
i. Prelogical	i. Mystical (Religious)
	Mythical
ii. Logical	ii. Speculative and Scientific
	Critical
iii. Hyper-logical	iii. Contemplative

It is hardly necessary to suggest the comparison of this scheme with the much-discussed "stages" of August Comte; the relation of our views to his and to those of others is elsewhere adverted to. It will be seen that Comte's second and third stages, the "metaphysical" and "positive," are both placed in our scheme under the one heading of "logical;" that his first or "theological" stage is covered in our "mystical;" and that to his three, thus made over into two, another, the "contemplative," is added. This last, as will appear below, becomes, in the sequel, the characteristic feature of our own interpretation, as given in Part III, below. In other words, the Positivism of the scientific view of the world is not the last word; a larger synthesis is reached in æsthetic contemplation and in its theory—the interpretation afforded by Pancalism.

PART II. THE HISTORICAL DEVELOPMENT OF INTERPRETATION

CHAPTER IV

EARLY RACIAL INTERPRETATION: ITS PRELOGICAL CHARACTER

IN the preceding chapter, the concurren development of individual and racial interpretations of objective experience of all sorts is pointed out. For the successive stages of the latter, the terms already found appropriate for the former were employed:[1] "prelogical," "logical" and "hyper-logical." Our further exposition, dealing with the racial progress of interpretation, will follow this usage; and the first great period, the "prelogical," will briefly occupy our attention.

§ *1*. *General Character of early Racial Interpretation.*

1. If the account given above of the relation between the individual and social motives be correct, certain pronounced differences between primitive societies and our own should strike our attention at the very outset; differences of a general kind which the later more detailed study should tend to confirm and define. If it be true that the primitive man's mental processes are largely prelogical, then so far as the individual is held strictly to social standards and interests, there must

[1] In an important work, to which reference is made below, M. L. Lévy-Bruhl also uses the terms logical and prelogical as applicable respectively to civilised and primitive thought (L. Lévy-Bruhl, *Les Fonctions mentales dans les sociétés inférieures*, 1910). See below in this chapter, sect. 6.

result a flowering of prelogical motives and interests unchecked, in great part, by the principles of logical organisation and validity.

Primitive man's understanding of nature and life we should expect to be of this nature. It should be shot through with emotional, practical, and mystic modes of apprehension. The sort of organisation found in the individual on the side of affective and emotional rather than of discursive logic, should be found at the fore. All the motives of an illogical and unlogical sort, called, from the point of view of superior logical development, superstition, fanaticism, prejudice, mysticism and self-contradiction, together with the motives arising from the social *rapport* itself, both individualistic and collectivistic, should come prominently into play. In the absence or relative subordination of control through truth and fact, in primitive thought, the motives of utility, personal safety, solidarity, tradition, fear, awe, prestige, would come on occasion to take each a preponderating place. The result would be, from our point of view, a state of confusion worse confounded; yet in itself organised in social ways. It is as if the child were released from the restraints and guidance of minds and sanctions more rational and consistent than his own; and were fed only upon illogical and emotional matter of apprehension and action. Such is, indeed, the place of the child in the primitive group; he is a member of a childish society. His thought can rise no higher than its source; he is under social compulsion to think and act as his fellows think and act.

This general fact gives us a sufficient indication of the marked differences we should expect the primitive interpretation of things to show from that to which we are accustomed. It forbids the application of logical criteria of truth and belief to the thought of primitive man. What is reasonable and normal to him may be irrational and abnormal to us. The only legitimate method of investigation is that of patient observation and sympathetic understanding of the primitive man's social life. What his mind is and how it works will then appear of itself. The most we can say is that there are no motives at work, however extreme their working may seem to us to be, that are not normal to the human mind taken as a whole. Different interests may dominate, but all of them are fundamental and natural human interests.

2. A second general consideration occurs to us when we consider the conditions of the rise of social life itself. No doubt the simplest group is the outcome of selection. The group seems to represent a unit in which collective utility, secured by some sort of co-operation and union of activities, replaces or succeeds the utility of individual capacity.[1] Selection acts upon groups and the fittest group survives along with the type of individual which belongs to it. In the animal world, this is seen in the utility of the instincts of collective life, by which the individuals are held to thc performance of acts of co-operation.

In man, the instincts, properly so called—the more finished actions of a species, performed without learning—are reduced to their minimum, and the group relies in their stead upon processes of education and training. The result of this is the imperative necessity for strict social regulations and sanctions; the very life of the group and its members alike is at stake. Hence there can be no abrogation or weakening for a moment of the

[1] This has been held by various writers. See the author's *Darwinism and the Humanities*, chap. ii.

forces of social restraint and direction—of social control exercised upon the individual—in the essentials of the common life. This is reflected in his willing response to the social demands. A social type of individual arises who is more and more imitative, obedient, and co-operative. The compulsion of social rule, like that of biological law, yields to the impulsion of inner motives of duty and conscience. Man is progressively socialised in the movement of his most intimate personal interests and motives.

3. If this be true, we should expect, the farther back we trace human culture, the more emphatic, dominant, and irresistible we should find the social means of organisation and control to be. Primitive man is governed by an elaborate system of rules, rites, and mystic observances which know no exceptions and show no mercy. We are accustomed to think of the "natural man" as a sort of primitive "individualist," free from our social conventions, and roaming at his own sweet will in the broad fields of life. But the very reverse is the case. Primitive man is a slave, subject to unheard-of severities, brutalities, terrors, sanctions, persecutions, all represented by detailed rites and ceremonies that make his life a perpetual shiver of dread, and a nightmare full of spectres. Nothing is so slight, not even his shadow or his dream, as to escape the regulation of the mystic powers, speaking in the social code; and nothing is grave enough to secure him a moment's respite or exemption from the penalties socially decreed. The savage is never gay; gayety is the product of civilisation.

§ 2. *The social Character of early Racial Interpretation*

4. All this is strikingly confirmed by the later anthropological investigations. The general expecta-

tion that early or primitive interpretation would be emphatically social and collective in its character, has overwhelming confirmation. Recent studies, notably those of the French school of sociologists, have brought this out quite unmistakably. The theory of *représentation collective*, worked out by Durkheim, Lévy-Bruhl and others, is based upon detailed observations, extending over a broad variety of primitive customs, rites and beliefs. Apart from the actual formulation of this theory and from its theoretical sociological interpretation in a doctrine of the relation of individual and society, we may accept the evidence as showing the common or collective character of the primitive meanings attached to things, and construe it in the terms of our own research in the field of "common" knowledge and feeling in the individual.

5. We may say, without hesitation, that primitive interpretation, considered as common meaning or *représentation collective*, is "syndoxic": that is, it is apprehended by the individual as being the common possession of the group, accepted by others as by himself. He makes no claim to have discovered or even to have confirmed it. It is a body of commonly accepted teachings—rites, observances, prescriptions, prohibitions, and so forth—for which he is not responsible, but which he accepts as being already established and binding. He is brought up from infancy in this body of syndoxic beliefs and apprehensions, just as the civilised child is reared in a body of socially recognised truths and usages. The difference is that at the logical stages of social culture, the individual comes sooner or later to criticise in some measure the social formulations, confirming or rejecting them in some detail, by the use of his own individual judgment. In this way the syn-

doxic becomes personal and "synnomic." This the savage or primitive man can not ordinarily do; he lives out his life under the domination of the collective tradition into which he is born. The social standards remain the only standards; the individual does not break through the social crust.

The reasons for this state of things will appear as we proceed with our analysis. Here we may signalise the absolute and compelling character of the elements of the social *régime* upon the individual, especially in the practical affairs of life. Society prescribes long and weary series of observances, giving no coherent reasons for the prescription; and the sanctions are the more severe for being occult, mysterious, and vague.

6. In further illustration of the collective character of primitive thought—especially as showing its pre-logical nature—we may utilise the results arrived at in our consideration of the common element in the individual's knowledge. We find certain striking points of analogy: points in which the two developments coincide or "concur" in the manner pointed out above.[1]

(1) In the first place the sort of necessity attaching to the primitive man's conformity to social rules is in general that which comes from a discipline due to sanctions external to him. By obedience he comes gradually into the social heritage of customs and duties. His spontaneous impulsions are trained in social channels, and his refractory appetites are curbed by rigorous

[1] In the preface to vol. iii. of *Thought and Things*, "Interest and Art," I have pointed out the striking agreement between our results, given in vol. i., and those reached for primitive thought by M. Lévy-Bruhl in his work *Les Fonctions mentales dans les sociétés inférieures*, a work in which the recent investigations of anthropologists are collected and interpreted. Certain points briefly indicated in that preface are more fully made out in the present text.

inhibitions. The penalties are those of an organised social order. They are frightful in their severity and unbending in their rigour.

This is only to say that society at this stage of organisation has not attained the results of individual reflective morality. The sanctions of right and duty have not replaced those of social convention based on prudence, fear, awe, and veneration. The active life is "syntelic," a life of common ends, common practises, common values, all socially rooted and sanctioned; it is not "synnomic"—not dominated, that is, by the individual conscience nor regulated by sanctions of an inner and moral sort.

(2) Another important character of this sort of collective meaning is its dependence upon the actual, the realised event or fact, as given in a present or past social situation observed or reported, and requiring nothing beyond this in the way of confirmation or proof. Once allow a certain sympathy between an animal and a man, and the animal is henceforth the "friend" of the man and his family. Given one fact following another, a mere connection in time or place, and a relation is established which is inviolable and mystical. The decrees of social convention are illustrated or justified by mere accidental happenings which, because not submitted to any sort of test, are just as open to one interpretation as to another, according to the trend of social interest and custom. Whatever is already accepted is found to be confirmed by new events; whatever contradicts this is non-existent, irrelevant. The actual then serves as sufficient basis for all sorts of preferential and selective interpretations, varying with the established states of mind and active attitudes of the group.

7. These characters of primitive thought find evi-

dent illustration in the existence and meaning of the totem. Differing among themselves as the theories of totemism do, we may still discover a certain basis of agreement as to the meaning of the totem to the savage himself. It is to him the symbol of his social group, the clan, the collective body of his relations natural and supernatural; this establishes the central fact of his life, his name and identity, by which and through which all other facts are to be understood. Apart from this group identity, he has no individuality. Facts, further, which in no way come into the sphere of the totemic influence, can not exist for him; but within the all-inclusive circle of interests which the totem represents, the rules, prohibitions, injunctions, sanctions, are absolute and final. Both in respect to authority and in respect to actuality, in giving what is to be accepted and in enforcing what is to be desired and avoided, in the theoretical reference as in the practical injunction, the totem is the syndoxic unit of social organisation, and with it of personal belief and practice.[1]

8. (3) Another approach to the syndoxic significance of prelogical racial interpretation is from the side of negation. Not only does the group assert and enforce certain positive rites and customs, it also established a body of prohibitions. There is the social "thou shalt not," as well as the social "thou shalt." Of course, in more developed society, where logical and reflective criteria are established, both in the intellectual sphere and in that of practice, the false and the wrong are matters of contradiction and opposition established

[1] The opinion is beginning to prevail that in totemism as found in Central Australian tribes, the most primitive type of culture is to be recognised. The recent work of Durkheim, *Les Formes élémentaires de la Vie religieuse*, gives a critical and comprehensive account both of theories and of totemism itself.

by more or less reasonable rules. "If this is true, that is not;" "if this is right, that is not." There are rules of logic, cognitive and affective, by which the negative establishes itself along with the positive.

But we have seen, in tracing the stages of development of negation,[1] that before the rise of logical negation or denial there exist very important modes of negative meaning. Rooted in active preference and selection, going back to original instincts and attitudes, there are strong movements of acceptance and rejection. There are motives of active rejection and of emotional repulsion in the individual, which do not arise from reflection or even from knowledge. There are rejections due to discomfort, to distaste, to custom, to novelty, to personal caprice, all determined by active and emotional motives, and some of them rising to the level of logical denial in their attempt to justify themselves.

9. There are two great classes of these active rejections. One comprises cases of positive exclusion, reached by a reaction of personality, through some one of its active interests, against the content or thing rejected. The child's rejections, all the way from a bad taste to an unwelcome guest, show this motive. It cxists in a great variety of forms in low and sublogical stages of mental process.

The other case is the mode of rejection known as "privation." In it the exclusion of what is rejected, incidentally to a positive selection or inclusive of something else. Such are the rejections of affection, sentiment, æsthetic appreciation, exclusive interest of all kinds. The absorption in the thing chosen and affirmed

[1] *Thought and Things*, vol. i., "Functional Logic," chap. x., and vol. ii., "Experimental Logic," chap. vii. *Cf.* also below, chap. vii., §§ 4, 5.

involves the exclusion and rejection of something or everything beside it. It does not intentionally reject a thing because of positive qualities, but only because of its non-inclusion in that which is selected for acceptance. The child's rejection of every doll but "Biddy"—however pleasing under other conditions the new doll might be—illustrates this movement of privation. The affection for Biddy is exclusive of any rival claimant to her place in the heart.

10. Admitting this distinction between prelogical rejection and logical negation, we find in the former a valuable clue to the understanding of primitive mind and life. The prohibitions, negative rites and observances, taboos, etc., of primitive peoples are very remarkable—more bizarre and unreasonable oft-times than the positive regulations themselves. The savage is hemmed in by a thousand rules as to what he should not do, what he should abstain from, in thought and action. The forbidden is forbidden under penalties of the severest nature. Most of what is allowed is still hemmed in by restrictions as to time, place, and circumstance, involving preliminary rites which are necessary to avoid disaster.

The remarkable phenomena of "taboo," and of the distinction between the sacred and the profane, must be looked upon as embodying the development of social rejections motived by emotional and active interest, either direct or remote; or as cases of privation, due to social absorption in other things and interests into which the thing rejected brings an unwelcome intrusion. Assuming, for example, in the later development of the religious motive, the belief in an exclusive and intolerant deity, we find, as a secondary result, the privative rejection of all other religious systems and objects.

Under the totemic system, matters determined in our culture by consanguinity, such as the limits of marriage, personal relationship of various kinds, and their negatives—such as the definition of incest—are fixed in view of the totemic emblem. Marriage between persons of the same totem is forbidden. These things are not at all matters of reasoning or logic; they are matters of positive social belief and preoccupation, fixed in rigid collective prescriptions both positive and negative. This means of course, in the first instance, the positive character of the social organisation of interest, of which these social rejections are an aspect; and we may note the fact that the rejections of the individual like his acceptances are socially regulated. The body of interpretation, negative as well as positive, is socially prescribed. The laws of social prohibition are collectively formulated, collectively sanctioned, collectively enforced; they are brought within the definition of the whole body of traditional interpretation as being syndoxic. But its syndoxic or common character is privative, not logical; it is of the sort that comes from affective organisation.

(4) Thosc points may be summed up by saying that primitive life is essentially religious; since it is in religious belief that the social and individual values and sanctions are alike conserved. It is the "theological" period of interpretation. The religious symbol stands, as we are to see later on, for the spirit of the community.

§3. *The a-dualistic Character of early Racial Interpretaticn*

11. We have noted the a-dualistic character of prelogical thought in the individual; the absence of the

dualisms which mark fully developed logical or discursive thought.[1] There are two important senses in which logical thinking, and the state of mind called reflective, are dualistic. There is, first, the dualism of "self and objects of thought," thinker and thoughts; this is the most refined form of dualism. It requires the conception of an inner life in which ideas or thoughts exist, and over against this that of realities to which these ideas or thoughts refer. This is properly called the "dualism of reflection."

But prior to this, yet persisting in the period of reflection, there is the dualism of substances, of classes of realities, the dualism of "mind and body." The progression of meaning by which this dualism germinates in early thought and comes to its maturity in the mode of judgment has been worked out in detail.[2]

Even the most casual examination of the thought and customs of primitive peoples shows that both these sorts of dualism, understood in any full sense, are absent. The material of experience is the same; but it is organised in other ways. For as we should expect, the absence of the forms of dual classification—mind and body, thinker and objects of thought—shows itself not merely in the simple absence of the familiar effects of such modes of division,[3] but in the freedom given to other motives which civilised individual thought hardly suggests to come forward with force and emphasis.

[1] For details, see *Thought and Things*, vol. i., "Functional Logic," chap. iii.

[2] See *Ibid*, chap. x., xi. The corresponding development in Greek philosophy, considered as culminating in the dualism of Descartes, is traced out in the writer's *History of Psychology*, vol. i.

[3] Such as belief in dead matter, the purely spiritual soul, the inner realm of experience, etc., all of which are late products of reflection.

12. In view of this, the various attempts which have been made to interpret primitive thought by certain more simple principles, such as association of ideas, animism, and so forth, have not been altogether successful. The fault of such theories consists either in making use of one term of a certain dualism, reducing it to a vague shadow of itself—as the "mind" term in the theory of "animism"—and attributing it to the savage; or on the other hand, in simply removing the logical function altogether from the primitive man's mind, and attempting to apply some simple principle of mental organisation, such as "association of ideas" or "reasoning by analogy." In all these theories alike, the assumption is made that the earlier stage of mental organisation, from which logical dualisms and classifications are supposed to be absent, is simply that which would be left if the logical functions had been subtracted from the individual consciousness as we know it. This results in a mutilated individual function, not a genuinely prelogical social one. How social thought organises itself, at periods at which the individual is not yet logical, and under the traditions of a social experience and habit still more primitive and un-individual, can be understood only by the actual investigation of primitive life. It can not be found in an individual brought up under other social conditions, however we may mutilate him. The working of the prelogical motives in the social setting proper to them must be allowed to show for itself.

13. The more positive side of primitive interpretation, due to the actual organisation of prelogical motives on a large social scale, will be taken up again below. Here we may note certain respects in which such prelogical thought shows the lack of the organizing

principles of discursive thinking: respects, that is, in which it is truly a-dualistic.

(1) The absence of the distinction between thoughts or ideas and actual realities—the dualism between the sphere of internal life or reflection and that of the external world in which things remain relatively constant—shows itself in certain striking confusions of primitive thought. It appears in the primitive man's inability to take a purely cognitive or disinterested point of view. The idea of an external system of things, existing in its neutrality apart from human interests and efforts, is indeed a late achievement; it has reached its full statement only in the positivism of modern science. But even its earlier form, in which it requires merely a certain discriminating perception and memory of things as such, apart from their value in practice, ceremony, and symbolism—even this is undeveloped in primitive societies. And this in the face of the compelling recognition on occasion of external things and maintenance of their cognitive value.[1] The savage denies the identity of an arrow with itself, refuses to call it an arrow, seeing in it a symbol of totemic rites and social values, which give it a different identity; but at the same time he seeks out the arrow and *uses it as such*, in the practical pursuit of game or in battle. It is as in the case of the child who refuses to admit that the doll is merely a thing of wood and paint, seeing in it the identity of a loved and cherished companion; but who also throws it about, sticks pins into it, and uses it generally under the practical admission of just the point before denied.

14. The reason of this would seem to be, in both

[1] This fact M. Lévy-Bruhl seems to overlook in saying that the savage can not isolate the object simply as perceived.

cases alike, the inability to isolate the content of apprehension as something which remains fixed, as just and only what it is, while the selective meaning or intent changes with the particular interest aroused. The separation of these two sorts of meaning from one another requires the recognition of the objective thing as something controlled apart from the selective and preferential use made of it, that is, apart from the subjective sphere of mind.[1] The dualism between the sphere of the external and that of the inner is absent. The former requires the beginning of the mediation of the real external thing through the idea which recalls and represents it; the other, the mediation of the further selective value or intent of the thing, through the use of the idea as means to personal ends.[2]

In the absence of this well developed dualism, we should expect just the sort of confusion that we find: the taking of a thing to be what, and only what, the dominant and exclusive interest, *as socially determined*, makes of it. On occasion, the interests change: one gives place to another, the thing takes on a different signification. The most curious results are produced in the domain of classification, where classes, due to emotional and active interest, take the place of those of a logical character that civilised man almost exclusively employs.

The result is striking enough; from the logical point

[1] In Greek speculation the point of view of the subjective as such, in distinction from the objective physical order, came only with Socrates, in whom the Sophistic movement culminated. But of course practical life among the Greeks, as everywhere, made use of the distinction—as does the savage and the child—without formulating it in a reflective interpretation. See my *History of Psychology*, vol. i., chap. iv.

[2] Both of these processes were introduced into Greek thought in the "relativity" of the Sophists. See *ibid.*, vol. i., chap. iv.

of view it is merely a sad confusion. The savage seems to fail to see, hear, feel, *straight*, to have normal associations, to remember correctly. One would take his imagination to be his only vigorous faculty, besides his intense and sombre emotions. He is haunted by fearful images. As already remarked, he is seldom careless; he experiences a wide gamut of social emotions, in the categories of which his perceptions are organised. A body of collective representation so assimilates his sensations that he does not isolate facts or judge in the light of what we call truth.

But when we recognise that a compelling and prevalent belief and custom, which does not separate the mere thing from all these accretions of social intent and sanction, absorbs him mentally, we see his position and understand something of his thought. He differs from the civilised child in this, that the latter is trained to become an independent observer and logical individual, to distinguish between what he sees and what he desires or prefers, to use the thing consciously as a means to personal and social ends, without denying its separate existence or distorting its objective meaning. All this the savage fails to get from his training; his teachers are as simple as he is. On the contrary, his training forces him to see only with the eyes of the social group, to hear only with its ears. The traditional mystical, religious, emotional reading of all his sensations and perceptions is given to him with his mother's milk; and in competition with the predominant interest of social solidarity, his motives to personal thinking get no chance.

15. (2) Akin to this is the other phase of dualism by which the life of reflection is clarified and illuminated, that between mind or life, considered as something

spiritual, and the dead material thing. Mind and body are terms of a substantial cleft in nature; not to distinguish the terms of this dualism is to lapse into evident confusion. This appears in striking form in early societies; and it is complicated by the peculiar interpretations placed upon animals in almost all primitive groups.

That this is in its nature prelogical appears from the absence of two great distinctions under which logical thought always labours: the distinction of persons from things, and the distinction of persons as individuals, from one another—especially of the personal self from other selves, of ego from alter. Both of these distinctions are further complicated in logical thought by the place assigned to animals, a matter which has been the occasion of certain interesting turns in the development of thought. In primitive thought, the place assigned to the animal is most suggestive. Considering primitive life, therefore, from the point of view of these two phases of dualism, let us take up the more developed first.

16. The radical distinction between mind and body is evidently not present in primitive thought; its absence is seen in the wide range of facts cited by anthropologists under the heading of "animism." These facts give unmistakable evidence that primitive man does not distinguish inanimate nature as such from animate beings, but apprehends the entire world of things as having dynamic and mystic properties akin to those of life and mind.[1]

[1] This is not, however, to accept the "animistic" theory—which in its traditional form tends to make the primitive man a dualist rather than the reverse, attributing to him a concept of mind which he is supposed to use to interpret body—but merely to utilise the facts cited, as evidence of the lack of dualism. To the positive interpretation of the facts we are to return again below.

Various facts show that the personal body is a source of difficulty and embarrassment, as we have seen it to be also in the development of individual thought.[1] The body is the locus of the most subjective attributes and activities, such as volitions, affections, and appetitive tendencies; but as a body it is, at the same time, a physical thing among physical things. In primitive thought, this embarrassment appears in the apprehension of all physical bodies as centres of both sorts of properties; and this universal complication of the two delays and renders difficult their later differentiation. The mind or soul is a thin vaporous body within the grosser person.[2] What appears to us as a confusion, however, is simply the recognition of things as they appear, an undifferentiated mass of phenomena which only later thought succeeds in distributing in dualistic categories.

A wide range of facts in connection with the rites attaching to physiological processes and functions—facts of birth, death, puberty, marriage, tribal initiation, personal cleansing and preparation for the chase, religious purification, and so forth—show to what degree the body is the centre of mystic a-dualistic beliefs concerning the nature and destiny of the soul. The many oddities of teaching which allow of the violation of spacial and temporal relations, the reverence for dreams and for the shadow of the person, the fear of physical monstrosities, all bear witness to the general fact, at least, that minds and bodies are not clearly distinguished.

17. Along with this confusion between the living and

[1] *Thought and Things*, vol. i., "Functional Logic," chap. v., sects. 5 ff.

[2] Early Greek thought retained this idea of soul as very refined matter. Probably not until St. Augustine did speculative thought attain the concept of a purely spiritual—non-material—soul or self. See my *History of Psychology*, vol. i., chap. vi.

inanimate—a difference which seems to us so evident and natural—the other sort of a-dualism, spoken of above as the confusion of persons *inter se*, is to be found, as it is also in the immature individual. There is the failure to distinguish "selves" as such—one's own self from others, "ego" from "alter." This confusion, as we call it in our dualistic enlightenment, is so deep-seated and fundamental that it colours primitive life and thought everywhere. The reason for it is the social character of the individual's training and discipline in self-hood and its responsibilities. Instead of personal identity and individuality, primitive societies develop a *collective identity*, in which the individual is merged. So radical is this that the stages through which the consciousness of self passes, as shown in the individual's development, are most painfully and slowly achieved by the race.[1]

Looking at this series of progressive stages, which need not be enumerated here in detail,[2] from the point of view of social organisation and growth, we find certain things worth noting in the racial development.

18. (1) The movement is not from the individualistic as such toward the collective; not at all. It is from an original impersonality, the primitive neutrality and lack of distinction of mere animation, towards a collectivity, which in turn undergoes various transformations in the progress of collective classification—totemic, religious, tribal, and so forth. The subjective and

[1] This appears also in the history of speculative thought. Theories of psychic "atomism" and monadism, or postulates of the isolation and impenetrability of selves *inter se*, do not occur in Greek thought; although practical individualism was sufficiently in evidence. And this in spite of the fact that physical atomism was clearly formulated.

[2] For details, see the writer's *Social and Ethical Interpretations*, 4th ed., chap. i.

living is distinguished from the objective and dead only in the development of certain interests; it does not take on the exclusive forms of logical classification. The determination of the single self finally over against other selves, comes only very late, when the logical and speculative interests are awakened. Before this in primitive societies the self is collective; it develops with the practical interests of the group, and its distinctions and oppositions are those proper to groups—families, totemic divisions, clans, and so forth. The meaning of individuality attaches to these larger units. The single person explicitly declares himself to be identical with his group, with his clan-totem, not with himself; and as other individuals are also identical with the same group, they all "participate" in a common collective identity.

This appears in the mass of evidence collected by the ethnologists,[1] which shows that the primitive individual does not and can not consider himself, even physically, a separate, distinct, self-identical being. He is one with his social class, through the unity of the self-meaning of that class. The material of the subjective comes from the group—symbolised often by an animal species—and returns to it again by the ejective processes of affective generalisation. In individuating himself, the savage individuates the class, the collective unit. Moreover, a great variety of inanimate things, comprised in the range of the totemic influence, may be included.[2]

(2) It follows that in respect to both dualisms—that of mind and body, and that of ego and alter—the germinating distinctions proceed in the interest and for the consciousness of the group. Whatever the col-

[1] Interpreted in the sense of the text by M. Lévy-Bruhl, *loc. cit.*

[2] Durkheim, *La Vie religieuse.*

lective usage and interest touches and claims becomes member and part of the collective self—persons, animals, utensils, weapons, dream personages, shadows, the dead, the mythical, the hero of tradition, the deity of the tribe. All these mean the *same self*, in the collective sense, and share in the mystical ownership and participation. The man is brother, other self, to the dog, the feather, the star, all because they and he are one in the collective life and interest of the common group-consciousness.

19. This extraordinary state of things is not disturbed, as we have already intimated, when the other dualism—that of mind and body—is also entertained. The hierarchy of interests and their relative integrity is illustrated. Each of the dead objects, to us so diverse in nature, although, along with man, a member of the group-self, nevertheless maintains, for the subordinate interests and ends of practical life, its separate identity also. For shooting, the arrow is an arrow; for cooking, the pot is a pot and the fire is a fire. Mere things are not such because they are determined by certain physical properties as not being persons, but only because they are determined as being temporarily or permanently excluded from other collective interests. The mere thing is the socially excluded, the non-tribal, the unmeaning thing.

This sort of distinction is, of course, more or less in the line of the coming logical classification of persons and things; since persons represent at the fullest the social values and interests, which are oftenest and most easily dislodged from the mere clods, the dead stocks and stones. But still the meaning of the distinction remains that due to the collective identity and interest with its correlative exclusions.

20. Both these senses, in which dualism in the logical form is absent, find illustration in the place held by animals in primitive life and thought. Authorities have frequently noted the utter unreasonableness of primitive attitudes toward animals and the complete disorder of primitive thinking about them. This obscurity is due not to the complexity of the subject, but to the embarrassment arising from the mystic and collective modes of apprehension of the primitive consciousness itself. Witness the totemic system as a whole and in detail. The animal may be a mere thing; as in the chase, slaughter, and consumption of game. Again certain animals—and any animal in its place—may share in the mystical meaning of the group-self. The interest invoked and the rites and ceremonies performed in the pursuit of that interest merge the animal with the man in the collective unity.

§ 4. *The relatively a-logical Character of early Racial Interpretation*

Coming now to enquire more closely into the positive nature of the organisation of social interests, by which primitive interpretation is constituted, we find certain outstanding facts which establish its prelogical character. Before noting the positive principles involved we may point out in what respects it fails to fulfil the positive marks of the logical.

21. (1) In the first place, it does not recognise the logical principle of "excluded middle." This goes with the fact that primitive negation, as we saw above, does not proceed by logical exclusion, but by positive and active social rejection. The classes established in this way, not being logically exclusive, may overlap one

another; the same "thing" may be in two or more classes, may be excluded at one time and not at another, may escape altogether the "either-or" of logical disjunction. The mutual opposition of classes, whereby a logical whole is exhausted, does not arise and the logical consequences of such exhaustion do not appear.

(2) This shows itself explicitly in the evident absence of logical contradiction in the thought of primitive peoples. They do not seem to be troubled by the demand we make upon our experience that it be consistent. They do not feel the need of rejecting a thing because they have accepted its opposite. Their canon of acceptibility is something quite different—emotional and conative satisfaction, the fulfilment of a social interest. This really arises from the state of things described above—the absence of a classification which requires mutual exclusion and exhaustion; for without opposing classes, contradiction in the logical sense can not arise.

(3) Many of the more superficial peculiarities of primitive thought arise from the same fundamental defect. The classes recognised are bizarre and obscure. Things are to the savage identical, which seem to us absolutely different and contradictory. Even numerical difference may constitute identity. The class is a matter of emotional and preferential value; a potency of harm, or of good, or of social utility, or of tribal advantage; such are the lines of distinction. The class is teleological, not logical. The interests are predominately active and affective; and the more exclusive social interests reduce the lesser personal ones to a lower and less important place.

(4) Hence also the lack of logical inference and argumentation. The force of logical necessity is absent;

the value of the universal is not seen. Instead, the savage cites mere juxtaposition of events or objects, analogy, the happening of omens, the power of rites and ceremonies—in all of which a certain compulsion of a dynamic and social sort is recognised. The novel is to him of supreme importance, for it reveals an unknown and real working of the power behind and within nature. The eclipse, the white man, the sudden attack of disease, the drought, these are omens of high significance. They give full play to the interest that is fulfilled in the realm of superhuman and mystical values. These agencies must be won over to the group interest, when to us they would be explained in the pursuit of the interest of knowledge.

Objectively, this appears in the methods of personal and social intercourse; discussion and argument give place to citation of ominous events and the invocation of social and supernatural sanctions. The processes of logical substitution and deduction are replaced by social conversion and confirmation, along with the appeal to established custom and belief. Conformity takes the place of logical community, the force of social constraint that of personal conviction. The motives of logical thought in general *are not yet released.*

With this more general statement of the superficial aspects of primitive interpretation, covered by the term a-logical, before us, we may now pass on to its characterisation. It recalls of itself the prelogical and affective organisation of the individual's interest.[1]

[1] Much of this chapter and the following (v) were presented in substance to the Psychological Seminary of Professor H. C. Warren, at Princeton University, in June, 1914.

CHAPTER V

EARLY RACIAL INTERPRETATION: ITS POSITIVE CHARACTER

§ *1. The social Organisation of primitive Interest*

ENOUGH has been said to show that the term prelogical has a definite meaning. Racially considered, it is a-logical in all the respects just enumerated. The further question arises in what directions other than logical the organisation of the interests of the social group takes on positive form? Is it analogous—and if so, how far—to the organisation of the individual's interests before the rise of explicit logical process?

1. It has already been intimated in what general respects we should expect racial interpretation, as socially constituted, to depart from the lines of individual thought. The individual, in civilised communities, lives in a group which has itself reached the logical stage[1]; its social tradition and its legal and conventional sanctions are reasonable in their intent, being cast in more or less rational form. Hence the individual, even when himself immature and prelogical in his thought, is constantly impressed and persuaded, educated and initiated, in lines leading on, by all

[1] The fact that in a great many things civilised society is ruled by convention and remains essentially irrational does not impair this general statement. Society can be logical when the reformer or moralist wakes it up to the necessity of being so.

manner of pedagogical short-cuts, to independent logical thought—or at least, to the semblance of it. Reasons, whether good or bad, are given for everything. The civilised individual, then, is not left free to develop the prelogical interests in their unchecked and spontaneous forms.

The savage child, on the contrary, is impressed in quite a different direction. He comes into a prelogical social heritage, as the civilised child comes into a more or less logical one. In him the full development of prelogical motives and social ends is free to show itself. The social interests and utilities, the social impulsions and sanctions, the social rites and observances—in short the social modes of organisation—dominate his mental development.

This leads us to expect that the racial type of interpretation will show marked characters due to the flowering of purely prelogical factors, which are not inhibited, but the rather encouraged, by the type of organisation already in force in fact and in tradition.

§ 2. *The affective Nature of Primitive Generalisation*

2. The first question to arise in this more positive enquiry is that of classification or generalisation. How does the primitive man, the primitive tribe, classify things and events? Does he identify a new object, for example, by its external marks, and classify it as belonging to a general class of objects already familiar? This would be to proceed along the line followed by the cognitive interest, the interest in objective truth.

It is evident that this is not the procedure of the savage. His interest in the new, as in the old, is not that of mere recognition or curiosity. His attitude is one of interested caution, fear, respect, awe. The physical properties of the thing open to his inspection are but its superficial marks; he expects it to manifest hidden and mysterious energies. Hence his interpretation takes account of accompanying phenomena, near or remote, of association with powerful agencies, of occult signs and omens. The position of the moon, the angle of the sun, the length of the shadows, the colour of the heavens, the flight of birds, the sayings of the oracles, the behaviour of sacrifices, the presence and character of dreams, the long-past event recalled—all these things give meaning and significance to the new thing. The generalisation effected is not one of mere cognition, but one of motivation, emotion, mystic participation, utilising the lines of association and knowledge merely as cues to indicate the direction in which the current of affective interest is to flow. The great casts of fate, the issues of life and death, the interests of tribe and family are the alternatives involved. The new thing has such a meaning; it is lucky or unlucky, favourable or unfavourable, friendly or hostile, with us or against us; its neutrality is a mere screen to its potencies of good or ill. The savage does not care for the mere thing; he wishes to know what sort of larger recondite mystic meaning and power the thing participates in, and what influence, fateful for better or for worse, its presence typifies.[1]

Not that he does this reflectively and intentionally; not at all. This is his spontaneous way of responding to experience. His emotional interest grasps the

[1] While recognising that religious distinctions such as those present in totemic classes, are of the highest importance because of their social import, still we need not go with Durkheim to the extreme of basing all classification, including the logical, on religion.

data of sense and organises them affectively. The tide of emotional interest flows over the mere fact and the presence of the thing becomes just that of its mystic meaning. While we, with our developed cognitive interests, naturally explore the thing, isolating it individually and freeing it of its ambiguities; he, with equal readiness, passes by its individuality, its detachment, and scents its intimate aroma of good or bad values, feels its participation in a class of powers that strike or bless.

3. This state of things, fully established by recent anthropologists, is readily explained in terms of what we have found to be the method of affective generalisation in the individual.[1] The method is the same. In the individual of our society, it is held in check and corrected as we have suggested, by the instruction, example, sanctions, of a social tide that sets in a different direction. But the same motives are present. In primitive society they have full opportunity to develop themselves.

The process consists in the formation, in the course of active and affective experience, of "emotional abstracts," active dispositions, attitudes, and moods. These take the place of cognitive classes; to one of these each new experience is assimilated, at the expense of its cognitive and logical relations. Emotional attitudes, active tendencies and habits, identities of value, take the place in the mental life of the similarities and identities of objective fact. These are general classes, in a true sense; but they have the force of general utilities, general values, selective and personal meanings, not in the first instance, at least, of general and impersonal sorts of fact, truth or existence.

[1] *Thought and Things*, vol. iii., "Interest and Art," chap. vii.

There is a real memory and recognition of feelings; and with this goes the generalisation of affective experiences in moods, sentiments, and active attitudes,[1] which take the form of personal preferences, interests, and tendencies. When by analogy, association, or suggestion, a new event excites such an attitude, an affective "general" establishes itself, giving meaning and class to the new experience. Thus a development is secured to the active processes of the individual, in relative independence of the growth of cognition and knowledge, but also is relative relation to it. The mental life is emotionalised, so to speak, where ours is intellectualised. Things are seen in the light of their mobility and potency, not in the light of their passivity and stability. The dominant interest is that of welcome or avoidance. The apparent and superficial properties of things become merely signs of hidden values, means to personal and social ends. The emphasis throughout becomes teleological; with the result that logical affirmations and negations, contradictions and exclusions, are insignificant in comparison with the acceptances and rejections resulting from the excitements, hesitations, and revulsions of the life of feeling and will.

4. All this is socially established in categories of tradition and custom. Its values are formulated in rites and ceremonies, its sanctions enforced by penalties of life and death, its forms rooted in the history of family, clan, and tribe, and cemented by the shedding of blood, and its heroes glorified in myth and story, the literary rendering of the corporate life. All the forces of the social organisation are enlisted on the side

[1] *Thought and Things*, vol. iii., chap. vi., and the literature of "affective memory" there cited.

of this mystical and emotional logic. What chance has thought as such to free itself?

§3. *Imitation and Ejection in primitive Thought*

Assuming that this body of mystical and emotional tradition is socially established, we may now enquire how it is socially maintained—by what processes of conversion and confirmation, as between individuals, and of common reading in forms of social prescription and sanction.

5. The establishment of great generalisations of conduct, emotion, and mystic observance, through social agencies and with reference to social sanctions, as just described, would seem on the surface to be a sufficient reason for their continued collective force and form. The individuals are all formed on social models, and trained in the same beliefs and observances. But we have found that differences exist between cognitive and affective modes of organisation in the matter of the establishment and maintenance of social community of thought. So far as the social matter, cast in tradition and current in social life, is formulated in cognitive and verbal form, each individual can absorb it through imitation and instruction. Thus the great body of formulated knowledge and belief is transmitted and conserved. There remains over, however, the more intimate part of the individual's conformities, covered by his personal feelings, motives, and attitudes as such, to which the representative forms of knowledge can not do justice.

This has been brought out in detail in our consideration of affective logic.[1] We have found that "generals"

[1] *Thought and Things*, vol. iii., "Interest and Art," chaps. vi., vii.

of action and feeling, representing repeated affective experiences, arise in the individual's mental life; but that some further process of conversion into social coin is required to secure their currency and social recognition.

For example, externally considered, the religious rites, being socially prescribed, become common property; and the literal fulfilment of these rites in common is easy enough; but the value attaching to their observance, as socially edifying, morally expansive and healing, as well as of personal satisfaction to the individual, can not be formulated. It can be symbolised, but in order to be symbolised, the meaning, however vague, must be real. There is here a mass of affective meaning, vital to the individual's observance and constituting the final motive, social no less than individual, which is in some way to be made available for collective use. Witness the undefinable but supreme mystic meaning of the totem. Even in civilised life, as we know, the essential spiritual values are in constant danger of being lost in mere verbal and social form. The spirit departs, leaving only the letter; the transitions of spiritual meaning may be enormous, while the external and representative form remains unchanged.

That this is true in primitive societies follows from the fact of progress itself. Progress represents repeated deviations away from rigid conformity, variations in the individual sense of value, which are in some way worked over into the body of social beliefs and observances.

6. Pursuing our comparative method, we think at once of the affective conversion which proceeds by the great processes of ejection and idealisation.[1]

[1] See *Thought and Things*, vol. iii., "Interest and Art," chaps. vi. and viii.

It is by "ejection" that the values of general feeling and habit are made social and collective, passing from the sphere of individual feeling into that of social community and unity of action, and secure the force of common consent and sanction. All this is shown in the remarkable series of facts upon which so much store has been set by the British school of anthropologists under the term "animism." Animism is a phenomenon of ejection; it illustrates the conversion of individual self-values into social meaning, the integration of private selves in the collective self. It is the leading motive of primitive morality, and of the idealisation which goes with it, in the great movements of early religion; to these, taken together, is due the essential mysticism of primitive thought.

§4. *Primitive Animism and Mysticism*[1]

7. The extent and variety of the phenomena of "animism" in primitive thought have been brought out with abundant detail by anthropologists. The hypothesis of animism as an explanation has itself taken on varied forms. What these have in common is the recognition of the tendency of the primitive man to interpret, or to act as if he interpreted, inanimate things as in some sense living or having minds. Nothing in nature seems to him to be dead, motionless, quite free from powers such as those which persons possess.

This has been understood by many anthropologists to mean that the savage, by some more or less conscious act, reads his idea of mind or life into things. Persons and living things are by right so understood; but in-

[1] Most of this paragraph has already appeared in the article, "The Religious Interest," *Sociological Review*, Oct., 1913.

animate things only by some sort of secondary process in which the true apprehension is supplemented by the tendency to "animate."

When this latter or secondary process is understood as explicitly following upon an earlier reading of the thing as lifeless, a dualism between persons and things is attributed to the savage upon the basis of which the animism proceeds. On this understanding, the opponents of the animists charge these anthropologists with attributing a complex dualistic process of interpretation to the primitive man. Whether this be true or not of any given writer, it is still important to bring out the fact that such a charge is not necessarily valid.

There are, in fact, no less than three different attitudes, on the part of primitive man, which may be described with some appropriateness by the term "animism," attitudes which, from the genetic point of view, it is easy to distinguish and estimate.

8. In the first place, the primitive "look" of things, before the distinctions of mind from body and of inanimate objects from living beings arose, would be in a certain way animistic *just by lacking these distinctions.* The cognitive interest not having been differentiated from the body of affective and social interests, things would not be apprehended as constant, regular, and law-abiding in their behaviour; but would be interpreted by means of the only mental classes then operative, those which represented the larger affective and active movements of consciousness. Whatever active motives and interests were operative, it is to these that the external data would be assimilated. Things could only mean what consciousness, at that stage of development, *could understand or intend them to mean.*

As a fact, this sort of interpretation, when present in the mind of the savage, would be shot through with emotional, active, and collective elements, and it would appear to the critical observer of primitive thought to be confused, mystical, and animistic. It is animistic in the sense that the first things of experience, the undifferentiated stuff of the inner life, would be charged potentially with the motives of a later dualism. The scientific observer is right in using the term "animism," but wrong in interpreting the facts in terms of a developed dualism. In view of its character as being first-hand and a-dualistic, we may call this first stage "spontaneous or a-dualistic animism."[1]

9. In the second place—and constituting a second stage—there is the sort of animism which arises through the processes of affective logic, that is, through the organisation of feeling, interest, action under social conditions. It issues in the general interests of the personal and social life and their conversion and propagation in the social body. The process is that whereby the mass of affective interest, embodying an interpretation of things and events in emotional and active terms, is subject to "ejective" reading generally. This is practically unhindered by cognitive discriminations such as those which the dualisms of mind and body and animate and inanimate would impose. "Affective generals," "emotional abstracts"—technical terms suggested in recent literature for the moods,

[1] The difficulty of naming it is the same as that felt, for the same reason, in the case of the early stages of reflection which preceded the dualistic interpretations of the world. Early Greek speculation was not really materialistic nor physical, properly speaking, but a-dualistic. I have used the term "projective" (*History of Psychology*, vol. i.) as a proper designation of this period, as preceding those to which the terms "objective" and "subjective" properly apply.

dispositions, and emotional habits thus formed—are passed back and forth in the give-and-take of social life, each individual using the common social interpretation to explain the acts and motives of others, and each applying the same rules to correct and justify his own acts and beliefs. This presents a very positive character to the sociological observer, who, not being aware of the emotional character of the process, assumes the presence of a conscious process of animism. But it is not animism in the sense that an idea, image, or notion of soul or mind is attributed by the savage to things; such a process is still to be developed. Let us then call this stage "affective or emotional animism."[1]

10. Finally, third, there is the full process of animism proper, arising with the development of the dualisms of person and thing and mind and body, and made possible only by these dualisms. These distinctions arise through the differentiation of experience into logical classes, which replace the emotional and affective classes of the earlier period. When this is accomplished, or is in the way of accomplishment, all sorts of ambiguities and compromises arise. The case of animism is that in which the new event is interpreted in terms of a conscious attribution of life or mind. It is no longer simply a spontaneous or ejective apprehension of the thing in emotional terms, a first-hand construction of the thing as living or personal; it is

[1] We find this in turn realised in the history of culture in the early pseudo-scientific interpretations of things which preceded true science. Astrology, alchemy, sooth-saying, magic, witchcraft—all sorts of "psychosophy," in short, to use a term of Dessoir's—are in this sense animistic: they assume the presence of occult, mystic, quasi-mental and vital forces. See Dessoir, *Umriss einer Geschichte der Psychology*, and the present writer, *History of Psychology*, vol. i., chap. ii.

rather a more or less reflective interpretation of it, following upon the apprehension of its positive marks or characters. This we may well call "reflective animism."

11. It has been said of certain of the "animists" that they make of the savage a philosopher. Primitive man is depicted as having an enquiring turn of mind, wishing to account for things, and asking for their causes; this leads him to the "hypothesis" that minds or souls lie back of the appearances of nature and "animate" lifeless objects.

This representation is true of the developed form of animism, the full process which takes place when the mind is sufficiently advanced in its cognitions and logical distinctions to begin the life of enquiry—to substitute curiosity for emotional interest. But the distinction of types of animistic process as made above aids us by way of suggesting a twofold elimination.

12. Recent criticism of the theory of animism makes it clear that reflective animism, supposing a process of conscious and logical animation, is not the true interpretation of primitive thought. The mass of the primitive man's apprehensions and interests are not personal and logical, but social and prelogical. He does not achieve, much less consciously make use of, the dualism of mind and body. He does not seek for causes in a speculative or purely explanatory sense. His interest, on the contrary, is in estimating the mysterious personal or occult powers of the new event or thing, and in adapting his life to them. His procedure, instead of magnifying and extending the discriminations of knowledge, passes them over without let or scruple, to arrive at the gain or loss their presence entails. His attitude is that of the interested participator, rather

than that of the disinterested spectator, of the course of nature.

We may therefore exclude the theory which assumes a process of conscious or reflective animation in primitive interpretation.

13. But it is as clear that we have before us, in the phenomena of primitive thought, something more than the mere chaos of unclassified experiences called above "projective" recognised by a second theory. In the interpretations of primitive man events are not by any means chaotic and lawless. He has his class distinctions drawn in lines of extraordinary precision and universality. To be sure it is not the lines of a logical classification; but still it is so definite that it requires of the youth a discipline of long duration, extreme privation, and lasting devotion to penetrate its intricacies and make ready for its requirements. The taboos, the rites of the chase, of marriage, of initiation, of forecasting, of augury in general—to cite only the well-known instances—represent the constitution of vested rites and liberties. Nothing could be further removed from disorganisation or mere disorder, whatever its superficial appearance.

Accordingly, we must disallow the "projective" theory: primitive life represents the organisation of a set of social and personal interests, its animism is an aspect of this positive organisation, not a sign of incoherence and confusion, or of the lawlessness of caprice.

14. It becomes plain, then, in view of this process of elimination, that the animism of primitive life is that of the affective type.[1] Its classifications are due to emotional interest, in its social rather than its individual form. We have seen that the body of collective

[1] The second of the stages described above.

"representation" is not at first strictly speaking representative; it is affective and conative. The processes by which in many cases it takes on representative form are those of a conventional and symbolic sort, by which the values of emotion and practice are socially fixed and made available in the life and tradition of the group. The totem,[1] the flag, the religious relic, the historical locality, stand for a lively and persistent collective meaning otherwise diffuse, intangible and unavailable to society.

15. The animism of the primitive man's interpretation of nature is indeed not one of ideas and thoughts, but one of emotion and practice: not one due to intellectual intercourse, to discussion and interchange of thoughts, opinions and proofs; but one due to the presence in individuals of emotional states which are socially organised by imitation, contagion, and ejection, and fixed by representative symbols.

The fact that in society there is a mass of established customs, rites, and habits which the individual naturally absorbs and observes—this sets the trend of his respect[2] and reverence, his disposition and practice, in social and conventional lines. His rational acceptance and belief follow after. He learns both what pertains in general to his companions and himself, and what pertains especially to each; what his status is, and what the rights, duties, and sanctions are that attach to it. All this becomes not merely "second-nature" to him, as we ordinarily understand that term—meaning acquired habit, over and above the natural formation

[1] This process, whereby the "emblem" or symbol comes to stand for community of emotional interest and social value is admirably brought out by Durkheim, apropos of the totem (*La Vie religieuse*, pp. 329 ff.).

[2] M. Durkheim (*loc. cit.*, pp. 304 ff.) adopts the word "respect" as I have, independently, for the believer's religious attitude.

of character—but *first-nature*, part of his very self, which is a determination of the social reality in individual form.[1] The social and emotional apprehension of things and events is original and fundamental; genetically speaking, it is prelogical.

16. The rôle attaching to the factor of ejection appears especially in the phenomena of animation. For the primitive observer both things and persons, animate and inanimate objects alike, have a selective and teleological meaning: things are intermediaries, agents, instruments of good or ill, of fate or fortune, or they are ends, beings to be propitiated, avoided, welcomed, appealed to, defended. In both cases they are values.

As such they can not be merely neutral things; they are always charged with forces and powers understood in analogy with those found in social agencies and in individuals. They are included in the interests that determine the preferences, dispositions, regulations, of society, and the loves, hates, fears, and revenges, which the individual feels in himself and finds stirring in his fellows. All this fund of meaning, semi-personal but social, quasi-subjective but of external authority, attaches to the things of nature as read ejectively by the individual. Things are centres of the feelings and motives he finds in himself; they are social fellows; like him they have interests, which must be given satisfaction.[2]

[1] I have suggested (*History of Psychology*, vol. i., London ed., p. 205), that if physical birth be placed at the beginning of the independent life of the child, that is, after the formative uterine influences have done their work, then personal mental birth should be placed, not at physical birth, but at the time when the individual becomes mentally a self or person, relatively independent of formative social influences.

[2] In the totemic systems, all the things of nature are divided up among the different totem-groups. Every tree, rock, brook, participates in one clan-life or another.

This gives in large part the colouring of mysticism of which we hear so much in the discussions of primitive thought and belief. By mysticism is meant, in this connection, just the absence of logical processes and principles, and the substitution for them of emotional and active attitudes and classes. There is more than this in primitive mysticism, as we are to see below; but this is the beginning of it.

17. But in this remarkable flowering of the emotional motives in the organisation of primitive social interests, the cognitive as such is not entirely lost. It is snubbed and kept under; but in many cases, not merely is there an attempt to give a reason, an account of the primitive rite or belief, but this account is often itself woven into the tribal tradition, and accepted as part of the intent of the collective meaning. Anthropologists have often mentioned this apparent need of primitive man to give some sort of rational, though fanciful, account of what he accepts, some reason "why." It takes generally the form of imaginative narrative. An association of an animal with a man in a dream, an event of a certain character following upon another, a remarkable victory won in consequence of the observance of a rite, a rival put to flight by a friendly animal, such incidents get all the force of confirmations of legend and justification of faith. The intellectual then follows on and gives support to the affective, a phenomenon by no means absent from more civilised thought. That this is the true order of the motives follows from the fact that any number of similar facts or occurrences that do not support the emotional belief are entirely overlooked or actually discredited. As with the mind-readers of to-day, any number of contrary or negative instances do not impair belief, while

a single favourable instance is cited as sufficient to support and justify it. When the traveller asks why the charm failed to work, or why the friendly animal did not come to the rescue, a purely accidental or personal reason is given for the exception to the rule. The animal was asleep, the god failed for the moment to recognise his friend, the moons were not observed. At any rate this serves to feed the dawning impulse to know. It also serves the end of securing objectivity and neutrality over against the mysticism of social participation and interest.

18. It is, therefore, by the intervention of the imagination that the reason gets a chance along with emotion and practice. It is analogous to that function of the schematising imagination in the individual by which he likewise interprets things and tries to explain them. In the racial movement it produces folk-lore and myth, quasi-reasonable accounts of things. These in turn take on social form, becoming stereotyped and traditionalised.

This constitutes, indeed, an essential step toward the logical mode of interpretation, since it supplies a fund of imaginative material which in time becomes matter of reflection. It is revised by the thinker who is able enough and bold enough to criticise it. The period from Hesiod to Homer is preliminary to the period from Homer to Socrates. In the child, there is the period of fairy-tale and epic romance, preceding that of prying curiosity and relative independence of judgment.

§5. *The rise of Mediate or Logical Interpretation*

19. On the whole, then, we have, in the progress of early racial interpretation, a picture whose outline is

familiar to us in the case of the individual. Two modes of apprehension, characterised as mediate and immediate are in evidence. The immediate mode is that of early projective knowledge: it is mystical, emotional, vaguely animistic, prelogical, holding sway from the start, but gradually yielding to the other mode which is truly cognitive. This latter takes on the forms proper to logical process, valuing objectivity and the related fact or truth, in opposition to social conservatism and the interested pursuit of collective ends. The cognitive interest aims at organising knowledge in an objective system of things, to which images or ideas become instruments or means of approach. Objects or things are mediated by ideas.

The transition to this latter mode is affected by the imagination, in its rôle of assumption, schematism, and heroic dramatisation. The immediate seeks to justify itself by the resort to objective and mythical pictures, thus adopting the weapons of representation. In this sense, the realm of social interests and values, like that of individual ends at the same stage of development, becomes intellectualised, and in so far satisfying to the reason as well as gratifying to the emotions.

This is the earliest form of the inquiry into the nature of things, before reflection is fully developed. It is the justification of beliefs already formed in the pursuit of other interests. But once introduced, it is the beginning of the mediation which goes on to be general to reflection, when both ends and facts are judged through the medium of ideas. Immediate and collective apprehension thus gives way *all along the line*, in favour of direct observation and logical proof.[1]

[1] See the further treatment of this transition in chap. viii., § 1, below.

There is, however, another fact of early social life in which the motives of collective conservatism show a remarkable organisation—an organisation so unique, so persistent, so complex, that its explanation serves as a sort of testing of theories of primitive interpretation. I mean, of course, Religion.

CHAPTER VI

THE RELIGIOUS INTERPRETATION

§ *1. The Religious Interest*

1. It accords with the method of our previous discussions to take up the psychological aspects of religion under the heading of "religious interest." By the term interest we mean to designate the larger mental dispositions in which the motives of experience and life organise themselves, in certain special directions, and terminate upon certain special classes of objects: the æsthetic interest terminating in art, the theoretical interest in truth, the practical interest in conduct and practical affairs. The inner organisation or interest, on the one hand, stands over against the objective content, on the other hand, to which the interest gives form and intelligibility.

Considering religion as being in this sense one of the great interests of mankind, found at all stages of racial culture, and everywhere in individual practice, the aspects under which it may be discussed—and is currently discussed—may be distinguished for convenience as follows: (1) The religious experience; (2) the religious object and its meaning; (3) the development of religious meaning: its logic; (4) the social character of religion; (5) the sort of reality that religion discloses.

These topics are treated in the paragraphs that follow.[1] The unifying thread throughout will be seen to be the genetic point of view by which the topic is made part of the whole of our work.

§ 2. *The Religious Experience*

2. The religious interest is universally admitted to be one that is, so to speak, "pointed outward"; it always terminates upon an object found by it to be holy, sacred, divine. Apart from this object, the presence of which it assumes, the interest as such has distinct emotional and conative qualities not found in the individual's attitude toward any other sort of conscious object. These qualities persist as substantially the same throughout all the changes that the religious object undergoes—from totem to Zeus, let us say—representing an attitude which is, for consciousness itself, *sui generis*. There is a true religious disposition, on the part of the devotee, believer, or worshipper.

This response in the life of interest is present wherever there is the suggestion of "sacredness"; it is absent when the sacred is absent. The secular does not excite it; and thc profane excites an opposing attitude of repulsion and avoidance toward that which produces a violation, injury, or profanation of the sacred. Just what the "sacredness" means will con-

[1] Much of the chapter has already appeared in the *Sociological Review*, Oct., 1913. Certain paragraphs reproduce in somewhat different form matter published from time to time in recent years in more special publications, *i. e.*, following the numbers of the headings given in the text: on (1), *Social and Ethical Interpretations* (1st ed., 1897); on (2) and (3), *Dictionary of Philosophy and Psychology*, art. "Religion (Psychology of)"; on (4), *Proc. Fourth Internat. Congress of the History of Religion*, Oxford, 1904, and *Darwin and the Humanities*, chap. vi.

cern us later on, in the discussion of the character of the sacred object itself.[1]

3. The components of this interest or disposition—the emotional and active factors which enter into it—are described by different terms which, however, on the whole, serve to bring out a fair agreement among the writers who use them. The early analysis of Schleiermacher, reducing the emotional part of the religious experience to feelings of "dependence" and feelings of "awe" or reverence, has stood as substantially correct despite various verbal substitutions in the description. Paulsen, seeking a more psychological term, substitutes "trust" for dependence; while theological writers, emphasising the intellectual, use "faith." Later sociological writers, seeing the matter more objectively, go back to the word "dependence." Likewise for "awe," different writers use different terms—reverence, fear, sense of mystery, respect, etc. But apart from these differences of verbal rendering, we may say that whatever the object may be toward which the religious attitude is taken up, there is the recognition in this object of a presence or force *worthy of respect* and *capable of giving aid:* it is "*auguste et bienfaisante.*"[2]

[1] E. Durkheim (*La Vie religieuse*, 1912) makes the distinction between sacred (*sacré*) and secular (*profane*) the fundamental one, but does not distinguish, so far as I can find, the secular from the properly profane; he seems, however, under the word *profane*, to include the secular (as in the application of the term to "matter," *loc. cit.*, p. 613). The profane, properly speaking, is by no means the same thing as the secular or non-religious.

[2] Durkheim, *loc. cit.*, p. 303. I cite in preference this new work of M. Durkheim, not only because it is recent and of high authority, but also because it is a work of detailed inductive research and of interpretation based upon facts (drawn from the totemic beliefs and practices of the Central Australian tribes). The full title is *Les Formes élémentaires de la Vie religieuse: le Système Totémique en Australie.*

4. In the individual, the religious interest always shows these two great aspects; and they both develop noticeably with the development of the consciousness of personality, that is with the apprehension of "self" and the correlative apprehension of other persons. I have elsewhere traced out this development in detail.[1] It preserves throughout—from the child's early physical dependence and appeal for help, to the high reflective sentiments of faith and praise for moral favours—its character as directed towards an object that is separate from the personal self, but in intimate physical and moral relation to it, and claiming a very special respect.[2]

As thus described, the religious interest, considered as a mass of organised human motives, reveals characters which mark it as being at once personal and social. Both the dependence and the respect which its object inspires are of the sort found peculiarly in social relationships: they are the same in kind as those directed toward persons. Religious dependence is not merely subjection to law, although its sanctions are external to the individual. Only in its crudest form, where it is a question whether, either in the child or in the savage, it should still be called religious, does the sense of dependence become mere fear, or religious worship mere recognition of authority or force.[3] Similarly, the feeling of respect or reverence differs from the sentiment of the sublime in nature, on the one hand, and from respect for moral law, on the other hand.

[1] *Social and Ethical Interpretations*, 4th ed., chap. viii., § 5.

[2] Spoken of more fully below.

[3] British writers generally, notably Tylor (*Primitive Culture*) and Frazer (*The Golden Bough*, and *Totemism and Exogamy*) refuse to admit that these sentiments are religious.

Moreover, the cult, in which these feelings take on active form, centres about a presence that sees, hears, sympathises, blames, punishes—that enters, in short, into personal relations with the worshipper. Its elements are prayer, praise, thanksgiving, and sacrificial, intercessory, and other rites.

In the system of activities, and in the institutions in which these activities are performed and developed by many persons in common, religion shows again its unique character. It is true that emotion and interest have generally their appropriate active channels of expression, either native or acquired; but in religious rites and observances, we find not a mere direct and, so to put it, "causal" way of reacting upon the world —a means of accomplishing an end directly, such as throwing a stone to hit a bird—but an indirect, personal way, acting by means of suggestion, information, persuasion, confession—all the means by which one person is commonly approached by another. It makes use of all the refinements of the personal *rapport*, from simple spontaneous suggestion, on the part of the savage, to the most refined and sophisticated logical subtleties, on the part of the learned theologian.

5. It is extraordinary, moreover, that in religion, both the privacy of our inner experiences of value, and the publicity of social sanction and authority, seem to be at their highest, although, perhaps, at different periods and with different adjustments of the motives involved. In the early prelogical religious interest of primitive man, the externality of the sanction is brutally evident; the religious authority does not differ much from those other modes of social constraint with which it is bound up, and to which

in many instances in itself appears to give rise. Here the element of privacy in the experience—the sense that the relation to God is special and personal to oneself—is at its lowest. Religion of this grade has been called "objective." Its rites, prohibitions, benefits, values, are socially regulated, publicly measured, and enjoyed by many persons in common.

6. On the other hand, religious experience goes on to become "subjective." The "religious consciousness" becomes the arbiter and judge in matters of belief and practice. The inter-subjective meaning of religion becomes that of a direct relation between the man and his god; and it loses something of the generalised or syntelic character so conspicuous in primitive religious practices. In the subjective religion of the logical consciousness, the religious allies itself with the ethical and in some measure absorbs its universal and synnomic force. But with it all, the religious relation becomes more individual and less collective. The flat "do not" of the savages' "taboo," and the equally arbitrary "do this" of the Ten Commandments, yield to the sympathetic "do this in remembrance of me," of personal union and communion.[1]

[1] It would seem to be at the two extremes of "objectivism" and "subjectivism" that those explosive and convulsive manifestations occur that give interesting variety to religious experience. The frenzy of the primitive religious dance is due to the public and collective suggestion and stimulation; as is also the fervour of the religious revival meeting. At the other extreme, the states of trance, ecstasy, prophetic mania, etc., in individuals, are instances of inner exaltation in which the communion with the deity and union with him lead to loss of personal identity and to pathological nervous manifestations. These "variations," treated interestingly by James (*The Varieties of Religious Experience*), are the accidents, so to speak, of religion.

§3. *The Religious Object: its Personal Meaning*

7. There are two senses in which the term "object" may be employed in connection with an interest or disposition[1]: it may mean either, on the one hand, the mere external thing, together with the symbols, ideas, etc., which describe or represent this thing; or, on the other hand, a further significance, meaning, or intent attaching to the thing, something symbolised or suggested. The distinction is a familiar one. Our interests always pour themselves, so to speak, over and around their objects and shape them into values, goods, or ends. The mere thing becomes a symbol or emblem of the larger object of interest; the "thing of fact" suggests and symbolises the "thing of desire" or will.

So far as the mere thing set up in the religious consciousness is concerned—the mere external symbol or emblem—it may be anything whatever. Actually it shows extraordinary variety in the different religions. A fetish, a totem, a churinga, a charm, a relic, a locality, a sound, an eclipse, an idol, a picture, things animate and things inanimate, may and do serve as objects of religious or quasi-religious veneration. It is clear, then, that it is not a particular thing as such, but some further value the thing possesses, that stirs the religious

[1] There are, in fact, no less than four different senses in which the terms "object" and "objective" have been used in this connection, and their distinction from one another is essential: (1) the mere objective thing set up before the worshipper; this may be properly designated the religious "symbol" or "emblem;" (2) the entire objective meaning or presence, the spirit, force, or god, which the emblem stands for or symbolises; this is, properly speaking, the "religious object"; (3) the cause, social, psychological, or other, which produces the religious experience in the devotee; this may be called the "cause"; and (4) the real existence or reality which religion may be held to disclose, discover, or reveal: the "religious reality."

interest. The thing becomes a signal, sign, symbol, emblem, of the further signification in which this value resides.

8. The thing suggests the religious meaning; and the question arises, what is the religious meaning suggested? It is a meaning, we may say at the outset, appropriate to the sentiments which the object excites —dependence and respect, both coloured in consciousness with a personal quality. The object appropriate to these sentiments must be either a person or something that suggests personality so directly that a relation is established to which that of person to person appears to be the only available analogy. Let us look a little more closely at these two cases.

9. (1) The first is the case in which the religious object is a god, a deity, explicitly described as personal. In developed religions the gods are personal beings.

All the great religions of the world have personal gods.[1] This was announced in Zenophanes' famous saying that all men's gods were in form like themselves. The limiting case, on one side, would seem to be found in the speculative and hyper-logical systems in which concrete personality is, for one reason or another, left behind or denied. In the systems of speculative pantheism and high reflective mysticism personality no longer appears. These, however, can not make good the claim to be religious, since the sentiments they excite are no longer pure, and the active rites of religious practice become altogether irrelevant. One does not expect personal consideration from the "universal order," nor does one worship "pure reason" or

[1] Buddhism (cited by Durkheim as an exception to this statement) was a religion after its founder was deified; whether it was a religion before that seems to involve just the point at issue.

"new thought." In passing from the religious state of mind to the theoretical and speculative, something is changed, the religious shading is lost.

10. (2) At the other extreme, there is another limiting case: the case of a stage of culture so primitive that religion, like other institutions, is reduced to its lowest terms. The religious interest is focused directly upon a physical thing, which seems to exhale no meaning beyond itself, no suggestion of personality. To this we return below; but here we may remark that the relation of person to person remains the only available analogy for the interpretation of the actual facts reported by the anthropologists. This appears both from certain positive considerations and from the negative one that no other hypothesis has the same grounding in facts.

The attempts to define religion in terms that exclude the personality of the object do so by citing the extreme cases mentioned. On the one hand, the object becomes an abstract principle. The gods of theistic religions become temporary, incomplete, or secondary embodiments of this principle, as in the "absolute religion" of the Hegelian metaphysics, or the impersonal force of the Stoic and Positivist theories. Religion is no longer a matter of religious experience; it becomes a category of absolute being, an effect of cosmic laws, or a postulate of social utility.[1]

11. The other extreme case, that of the primitive cult, is cited principally by sociological writers, who in their search for the root-motive or simplest form of

[1] Of course a metaphysical theory of religion is legitimate in its place, but it can not take the place of a definition of religious experience. Religion as cause or effect, as category or postulate, is not the religion of the devout state of mind. That which causes religious fervour may not be at all that towards which this fervour is directed.

religion, seek for some character of sufficient generality to include impersonal as well as personal objects in the religious interest. A typical attempt of the kind is that which makes the mark of religion the "sacredness" by which the primitive man distinguishes the religious objects from other things: sacred things and events are religious, secular things and events are not so; there is a fundamental cleft between the secular and the sacred, extending through the whole of nature.[1] Both sociology and psychology, one citing anthropological facts and the other the psychological manifestations of religious interest, demand an account of *sacredness.*

If we are to escape the mere tautology of saying that sacred things are religious because religious things are sacred, we must give some actual and distinctive content to the concept of the sacred. The task of both these sciences alike is to discover this distinctive content or meaning. Given the same material thing, why is it sacred in some circumstances and in other circumstances secular or profane?

The task assumed by the advocates of the theory just cited is to show that in some cases sacred things do not have the suggestion of personality, although in such cases these things are the object of a veneration or respect truly religious. The latter part of the theory is essential, since even though lacking personality,

[1] This theory, announced some time ago by M. Durkheim ("De la Définition du phénomène religieux," *Année Sociologique*, ii., 1897), is developed in his work *La Vie religieuse*. In order to sharpen the issue, we may observe that, in English usage at least, sacred is a term of the widest import: all things subject to any degree of religious respect are "sacred" (such as the vessels of the altar), while the gods themselves are not only sacred but "divine," and persons (especially priests, saints, etc.) and places most intimately associated with the gods or with religious ceremonial, are "holy." The sacred, therefore, is the most extended but the least significant, of the three terms.

such things may prove to be merely the symbols or emblems of a further significance in which the suggestion of personality does enter.

12. It is just this latter requirement, the second part of the task, on which this theory suffers shipwreck. It has no difficulty in showing that in primitive religions, of the totemic grade, for example, objects in themselves quite impersonal—the thuringa, the graphic sign, the external thing bearing the totemic name, etc.—are held to be sacred. But it then fails to prove the further point requisite to the definition: the point that the entire meaning upon which the religious interest terminates is devoid of personal suggestion or intent. Either it fails to subject to an adequate psychological analysis the notion of personality at this stage of culture, or it fails to state positively what the actual impersonal meaning of religious sacredness is. In both of these directions an adequate analysis serves, in my opinion, to show that the religious object, even in totemism, is not without the marks of animation or life, and that it is, in a crude sense but still positively, *quasi-personal for the consciousness of the devotee himself.*

13. As to the meaning of the totem—it symbolises the clan, as all authorities agree. The important members of the clan, the true members, are persons. After the persons of the clan, come particular animals (the totemic animal or plant itself being a species, not a particular one or more), and finally the things found in the totemic area. The true reality of which the totem is the emblem (akin, in the words of Durkheim, to the flag in civilised countries) is the human or, at most, a human-animal group.[1] Now in what sense

[1] In which (Durkheim, *loc. cit.*, p. 366) mythical ancestral personages also figure.

can this group, as distinct from its individual members, be called impersonal? Is it true that, in becoming the symbol or emblem of a group, the sign, itself generally a living species, acquires a meaning from which the essential attributes of the members of the group are omitted? If the savage, who probably does not think in terms of logical classes, but of affective and social groupings, were asked what the totem was, he would point no doubt to individual persons or animals that shared the totemic name and "participated" in its meaning.[1]

No doubt the constraint exercised upon the primitive man by society is not attributed by him to single members of the clan; but in locating it in the totem, does he give it an altogether impersonal representation? Rather we should agree with Durkheim in ascribing to the totem a generalised vital force. If so, the conclusion seems to follow that it suggests the animate, as an essential part of its meaning, and that it is therefore—as we are to show—in so far crudely personal. That is to say, it can not be properly described as "impersonal."[2]

14. It remains to make good the statement that the theory in question, which is typical in this respect of the theories of primitive religion hitherto in vogue,

[1] It would be difficult to maintain that, even to us with our logical notions, the concrete flag floating before us is entirely impersonal in its meaning. As a social emblem, it stimulates just those collective emotional dispositions that lie at the root of our sense of personality. The flag-meaning, apart from the mere number of flags denoted, is an affective and social, not a logical class.

[2] In the article cited, *Sociological Review*, Oct., 1913, further criticism is made of M. Durkheim's position. In a personal note to the writer commenting on that article, M. Durkheim says that by "impersonal" he simply means "collective." This is not, however, the meaning usually given to the term.

fails to give thorough analysis to the meaning of personality.[1] Without this any theory is incompetent to decide whether or not a given symbol or emblem, such as the totem, possesses the meaning of personality.[2]

15. A true statement of religious animism would recognise a development of the meaning of the soul-principle which the savage ascribes to things, from very crude beginnings. This principle passes from mere vital attributes up to the substantive and spiritual soul, as has been shown in detail above.[3] Analogy drawn from the present-day genetic psychology of self-consciousness not only allows but requires this. It would be impossible for the savage, whose representation is prelogical and whose interest is affective and collective, to isolate a "soul," as we understand the term, in the sense of something individual and spiritual, in himself or in anything else. He has no such "idea," no "notion" of the kind. But this fact does not refute the animistic interpretation, nor require us to resort to a "pre-animistic" stage of religion. It does not even justify one in calling the totemic system non-animistic, whether or not we consider it religious.

For while the savage has no idea of soul as something different in nature from body, he still has a feeling, an interest, a collective intent which, so far as

[1] It is a current criticism of the British "animistic" school that their psychology is out of date and untrue. They are for the most part associationists, who find in the sense of personality a "notion" of the self, an "idea," having always the same meaning (see Lévy-Bruhl, *loc. cit.*, and the writer's *History of Psychology*, vol. i., London ed., p. 16).

[2] Writers of the French school in turn deliberately court this criticism in resorting avowedly and almost exclusively to the objective or sociological point of view, and in showing a certain inhospitality to the psychological. See the citations made in the article mentioned (*Sociological Review*, Oct., 1913, p. 9).

[3] Chap. v., § 4.

it is developed, is the way he apprehends the mental and animate, even though he may not "represent" it at all clearly.[1] He probably has no notion or idea of "inanimate" in contrast with animate, much less of body *versus* spirit, although, in his practical adaptations and social rites, he makes the actual discrimination of living and not living.

16. The principal characteristics of the primitive man's view of things are such as to justify the application of the term "affective animism."[2] There is a recognition of things, animals, persons, as possessed of a certain forcefulness, a dynamic quality akin to animation. It would not do to say that things are taken to be "living merely," for "living-merely" is too high and abstract a notion. It supposes the contrasted idea or meaning of something more than mere life, a sort of spiritual power. But we may say that to the savage the living shows the presence of those powers which become later on, in the more mature thinker's reflection, the signs of the spiritual. To the primitive man, the one dynamic centre stands for all the confused mass of meanings which are still to be differentiated into the categories of inanimate,[3] living, and spiritual. But if the researches of anthropologists should reveal a culture still more simple—one quite a-dualistic and projective, in which experience is near the grade found in the perceptual intelligence and

[1] He represents it generally as a small material body, or a second more refined physical shape, or as something resident on the breath—all natural hypotheses taken up in the early stages of Greek speculation.

[2] The second of the stages described above, as has been said.

[3] The case of the inanimate first clearly distinguished by the savage—under the totemic régime—is probably the dead body, just after death. He thinks the soul has departed from it, leaving it in this condition. It is different from the things of nature, for they still retain their mystic forces, while the dead body has lost them.

instinctive action of the higher animals—I think we should still have to recognise a strain of animistic meaning running through it: an animism of the first or "spontaneous" sort. For certain of the animals, though lacking the distinctions of human thought, still appear to apprehend the dynamic quality of nature, and to look upon movement and change as being the source of the same type of experiences as that which they derive from man and their fellow-animals.

To the child, also, even the most original panorama of the external, presented to the gaze of the mind, would not be a flat motionless sheet; but a scene of change, explosion, colour, vitality. And in it, moreover, by the very conditions of its perception, the vague lines of differentiation into parts would appear, which the dawning instincts, appetites, and interests would rapidly produce. It would be like the segmenting egg, in which the lines of cleavage, division, and partition, prophetic of the coming development, are just beginning to show themselves. The *Anlagen*, to pursue the biological figure, are present: the distinctions of life, mind, and things in their embryonic form. This would appear to the outside observer, to whom the mature distinctions were clear, as a confused and mystical mode of animation.

17. The theory, therefore, that denies all suggestion of personality to early religious objects, the totemic in particular, fails at two essential points. It recognises the symbolic or emblematic character of the mere thing, the totem or idol[1]; but in admitting that the

[1] The mere superficial fact of "sacredness" attaching to a thing means nothing. The sacredness of the inanimate thing is reflected, not original. It is sacred because it symbolises a meaning beyond itself—vitality, personal force, God. The fountain of sacredness is to be found in that which is symbolised.

religious meaning proper resides in the further suggestion of some sort of force or life, it allows an animistic interpretation. Further, the theory fails to show, by an analysis of the primitive man's representation and interest, that such an animistic apprehension is not a stage in the development of the animism of conscious personality.

We are able to say with confidence, seeing the results of genetic psychology, that it is such a stage, a necessary stage. If as psychologists we were called upon to construct in advance the sort of world-view primitive man as we find him—prelogical, collective, mystical in his intuition—would entertain, it would be just this and nothing else. His gods could not be spiritual, individual, and intelligent agents; they could only be vaguely animistic, dynamic, collective, mystic presences, satisfying to the crude quasi-personal interest by which they are apprehended.[1]

§4. *The Religious Object: its Ideal Meaning*

18. In the current treatment of the ethical and æsthetic modes of experience, the characters attaching to ideals are much discussed. Both moral right and æsthetic perfection are ideals; and the religious object shares the ideal quality with the moral and the æsthetic.

What this means will appear on a closer examination of the object set up for worship in this or that religious cult. As we have seen, the mere thing is not the object; it is the symbol of the object. There is a further meaning, a signification beyond the mere symbol; and this signification is one determined, not by representa-

[1] As I have said elsewhere, if the child, at a certain age, were asked the abstract questions the anthropologist puts to the savage, his replies would show the same puzzling confusion and incoherence.

tion or thought, but by emotional and active interest. It is not an associated idea, or a system of ideas, but a worth, a further value imputed to the concrete object. The sacred thing is the centre of worthful experiences; the deity is well- or ill-disposed, capable of harm or benefit. All such unaccomplished, unfulfilled worths, presenting something to aim at, to live for, to desire, are ideals.[1] In this sense the religious object is an ideal.

19. But more than this. The religious object is always, as we have seen, endowed with the attributes of life and personality, however vaguely. It is a more or less developed centre of inner life and spiritual force. If this be so, then its ideal character must be that attaching to such an object: it is an ideal of personality, of what the worshipper himself is, a self in some more or less adequate sense. In what sense, we may ask, and by what process, can a meaning of selfhood or personality take an ideal form?

In the development of the ethical ideal in the individual this movement is seen in operation.[2] The growth of the individual self-consciousness proceeds by an organisation of factors which are, so to speak, prospective, forward-reaching, active. We have within us the actual self of habit, ready at any time for action, which finds, however, over against it a self of accommodation, learning, imitation, adaptation. A better self—more

[1] See *Thought and Things*, vol. iii., "Interest and Art," chap. v., §7; also vol. i., chap. xix, §8. This worth is embodied in the imagination, by which the object is schematically developed and its possibilities brought out. The ideal is not a model of perfection, brought into the mind from some outside source; nor is the process of idealisation a special and mysterious faculty or intuition; it is simply the imagination in its assuming, schematising, and experimental rôle, running ahead of knowledge.

[2] *Cf.* the author's *Social and Ethical Interpretations*, 4th ed., chap. i.

adapted, better informed, more obedient, virtuous, just, generous, wise—hovers before our gaze, and serves as model, inspiration, imperative, and ideal. At the limit, this self is perfect, the all-good and the all-wise, the ideal Self.

The religious ideal is of this character. But it is not the same as the ethical ideal; and the contrast between them is very instructive.

20. The ethical ideal is felt to be *ideal*—that is, *not actual*. It is a norm set in the mind, a rule of ideal conduct imposed upon all persons, but not realised by any of them. We do not suppose our moral ideals to be actually fulfilled, except in God. This is to say that the personal ideal *qua moral*, if it is to be realised, must become also the personal ideal *qua religious*.[1] It is, then, the religious ideal proper that is actually realised in God, not the merely moral ideal. Our moral nature postulates an ideal ethical value, but not an ideal ethical person.

The same thing appears also from the point of view of duty. The moral ideal is one of duty as between persons. It is a rule of social intercourse as well as of personal perfection. It could be realised in practice only in a society of perfected beings, a utopia, which does not exist. Such a society has been dreamt of and written of, but never seriously postulated as existing, except in the realm in which the religious ideal is also imposed: the kingdom of God, the heaven of various religions, the state of perfection of religious mysticism. On the side of society, therefore, no less than on the

[1] It is on the basis of this truth that the historical argument of Anselm for the existence of God can be reconstructed. Instead of an idea whose infinity stands in the way of the omission of actual existence from its object, as argued by Anselm, we have the postulate of the ideal self, whose integrity demands its existence.

side of the individual, the postulate of an actually existing ideal order of persons is not moral but religious.

21. Despite these differences, however, the two ideals are closely related. Their common trait is equally fundamental to both: the relation of the personal self to another self viewed in the light of ideal personality. In the ethical, both these selves are concrete, actual, social fellows, existing side by side; but the ideal of their relation does not exist save in their thought. In the religious, the second or "other" person, the deity, is one with the ideal, which is thus made actual, brought into human life, and symbolised concretely in the religious emblem that stands for God.

Thus understood, certain positive marks attach to the religious ideal.

(1) It is the fulfilment of the entire personality, not merely of this or that function or capacity of the self. The savage finds the power behind the movements of nature mysterious, awful, because he cannot anticipate or discount the resources or decisions of the Great Spirit. So the voice of Jehovah commands, "Take off thy shoes from off thy feet, for the place whereon thou standest is holy ground." The creeds of theology, which ring with terms of infinitude and juggle with synonyms of personality, are as inadequate to express the religious ideal, as are the groans of the terrified savage, or the genuflections of the awe-stricken pilgrim at the holy place. It is here that religious awe—a singular combination of fear, respect, and admiration—meets its object half-way. As perfect this object is admirable, as personal it commands respect, as powerful it is to be feared. But all of these attitudes are coloured by the glamour of the ideal. It is not ordinary fear, but fear of one having endless power; not ordinary

respect, but respect for the morally highest, the supremely excellent; not ordinary admiration, but acquiescence in eternal wisdom and contemplation of divine beauty. These feelings are fused in one sentiment, one interest, as in the object the attributes are fused in one ideal person, God.

22. (2) In the religious interest, the social relation takes on a form analogous to that of superior and inferior in our daily life. The ideal of personality which we enforce upon ourselves as a moral ideal, we attribute to God as his by right of actual nature. Although he is one of the persons in the social relation, being divine he is ideal as well.[1]

In presupposing the actual existence of its object, however, religious faith generates a relation of great import and real effect. God is an actual presence, like that of the child's superior—father, mother, elder brother, counsellor, friend—all designations realised in actual religious creeds. These are not mere figures

[1] Now that we have discussed the religious object, we are able to detect more clearly the elements of the religious emotion. Religious "awe" is complex: it contains *fear with admiration* (*l'effroi dominé par l'admiration*, Leuba, in *Revue Philosophique*, October, 1913, p. 413). Mere respect too, even though personal, is not all, nor veneration (as due to excellence, age, experience, wisdom, etc.); it is *admiring respect*, æsthetic absorption and appreciation akin to the sense of the sublime. This is seen in the historical bond between art and religious technique. The intellectual emotions as such—curiosity, logical satisfaction, etc.—are very subordinate, especially in the primitive stages, where religion is prelogical and social. The savage's curiosity, so-called, in the presence of the religious symbol, omen, or portent, is more of the nature of an alertness or attentiveness, to get the first sign of the will of the god, similar to that of a player for high stakes, who watches the turn of the dice. And the theoretical interest as such, prominent in systematic theology, is rather a reflection upon religion than a direct participation in it. The dogmatic development of theology does not in itself involve personal religion; in fact, this latter involves dropping the critical and logical attitude of mind.

of speech, but aspects in which the actual presence is symbolised to this believer or that, in this emergency or that. Through this personal communion the individual finds his forces renewed, his courage revived, his emotions purged, his aspirations directed, his visions of beauty and good clarified.[1]

23. The dependence felt in the religious consciousness is therefore of the nature of dependence upon a superior person. It extends to all sorts of aid or succour, physical, moral, intellectual. And the superior person is one to whom the relation is also important. The dependence is in a sense reciprocal. In primitive societies, man protects the totemic animal, as the totem protects the man. In religion everywhere the sacred places and objects are committed to the guardianship of the priests; and the sacred truths are defended and enforced by the prophets and law-givers. The gods are represented as jealous of their rights, gratified by praise, pleased with offerings, and given to pursuing their own interests or glory[2] through human agencies. This means that the significance of the relation is not unilateral, so to speak, one to which one member is indifferent. On the contrary, the deity of the tribe finds his life and interest in those of the tribe; he pursues with unerring vengeance and direful penalties those who go out "after strange gods." He chooses a "pecu-

[1] This is in contrast with the discouragement which comes from meditating upon unrealised ethical excellence and unperformed moral action. The religious ideal stimulates and inspires. It is no doubt this somewhat hidden motive in the whole that develops itself by making concrete the person of God in an incarnate form, as saviour, mediator, virgin mother, saint, etc. The need and the effect of the "Imitation of Christ" are very real; it means the absorption of the ideal by communion with it.

[2] As in the Westminster Confession of Faith.

liar people," who become the special instruments of his revelation as well as the special wards of his providence. They represent his interests.

24. In the more reflective stages of religious development, this concrete anthropomorphism tends to disappear, just as it itself follows upon the cruder animism of the primitive cult. The ideal takes on more and more refined form. But it never becomes an "idea of the infinite," except in our logical rendering, nor a formula of final perfection, except in our theory. It remains a feeling forward, a prospective intent, a sort of drift or momentum in the actual organisation of interest, toward the fulfilment of the full promise of personality. Hence the inadequacy of all symbolism, as well as the futility of all logical statements, of the nature of God. His nature remains always personal; but the sense of the sort of person he is changes with the development of man. The human ideal is a restatement always of the human fact.

§5. *The development of the Religious Meaning: its Logic*

Coming to ask as to the processes by which the religious interest develops, carrying the religious object along with it, we find certain striking complications.

25. In considering the development of the cognitive function—its genetic logic—psychologists point out the familiar processes of memory, representation, conversion, generalisation, etc., through which a given content or object normally passes. And in considering similarly the logic of emotion and interest, other processes are discovered more or less analogous to the

former, but also having certain novel features.[1] In the latter, affective generalisation, ejection, and idealisation, are the mental functions most in evidence; they take the place held in the development of knowledge proper by the cognitive processes mentioned, which culminate in conception, judgment, and reasoning.

26. Now in the religious life, we find a singular union of these two great modes of development within the mind, the logical and the teleological, as we have called them respectively. The logical erects classes and establishes facts and truths, by its methods of proof; the teleological issues in affective interests and defines ends and values. Now, in the religious life we find the object, God, looked upon as *really existing, as if* established by processes of knowledge, while, at the same time, it is determined by the religious interest *as an ideal or end.* Religion claims to present both a system of truth and a system of personal and social values. God is *both fact and ideal;* not merely in the common way of a value attaching to a fact or truth, as utility attaches to my inkstand, but in the peculiar way in which a meaning attaches to that which symbolises it. The ordinary attributed values are real only in so far as the things in which they inhere are real; when unrealised they are ideal, and not actual. But the meaning of the divine, the sacred, the holy, is not "attributed" to the object in this sense; on the contrary, the objective symbol or emblem is attributed or assigned to this meaning, to represent it vicariously. The ideal exists and gives a new sort of reality to the thing taken as symbol.

[1] The latter, the affective logic, has not been fully worked out; the author's contribution to it is to be found in vol. iii., "Interest and Art," of *Thought and Things* (the logic of cognition being treated in vols. i. and ii.).

27. As to the logic of religious interest, we may say that it is that of interest generally. It seems to be at first collective, traditional, socially propagated; the single person being the mere channel of its expression, a *locus* at which the personal forces are moulded into shape as the vehicle and instrument of the social will. The gradual refinement of the religious interest proceeds with the growth of the individual self in competence and independence. But it is just in the religious interests of mankind that the traditional forms hold their own most conservatively. The social processes of imitation, contagion, ejection, constraint, dogmatic formulation of creed, all serve to increase the rigidity and solidarity of religious belief and practice. Myth and folk-lore are the antecedents of doctrine, mystic ceremonies precede the more reasonable rites of worship. A legitimate glamour is thrown over the whole by the elaboration of form in architecture, painting, and music, through which the motive of æsthetic admiration is developed. We see here, in short, all the processes of affective logic exemplified: the matter of the religious interest is affectively generalised in dispositions and modes of actions; it is ejectively converted from one mind to another and propagated through the established institutions; it is embodied in the highest æsthetic products of human art. The church presents the essential positive features of religion and morals, as well as their limitations.

§ 6. *The social Character of Religion*

28. The result of our inquiry, so far as we have now gone, is in striking accord with those of sociology and anthropology. The collective character of the religious

interest, as embodied in all that goes to form a cult or church, is universally conceded. But the newer researches in primitive culture show further that social organisation itself, even in its most secular details, has been dominated by requirements and distinctions rooted in religion.[1]

This fact has been utilised in different ways in theories of religion. Comte indeed anticipated the point, by recognising a stage of "theology" in the general evolution of thought, a stage at which the religious point of view was adopted to explain the world generally. At this stage, man was still under the domination of the religious interest; his interpretation of things was theological. Only in later stages of the development of culture, the "metaphysical" and "scientific," was he to free himself. But not entirely, for in the last period, that of "positive" and scientific thought, the need of religion was to find its satisfaction in the concept of humanity, to which an altar was to be erected bearing the inscription, "Religion of Humanity." Man conserves the religious interest, therefore, but he contents himself with a very platonic friendship for God, taking on the form of a certain flattery of himself!

29. Another theory, starting out from the same social presupposition—that of the union, in early times, of the collective and religious interests—suggests in its conclusion a different alternative. Humanity outgrows religion, we are told, in becoming scientific and positivistic. Hence our seeming progress towards the "non-religion of the future."[2] This shows itself in the gradual secularisation of all our interests. There will

[1] The totemic system of group organisation is a very striking instance of this, as different writers agree.

[2] Guyau, *Non-Religion of the Future.*

remain no divine, no sacred, even no profane—only the secular.[1]

The facts recognised by sociologists to-day, however, confirm the conclusions reached above, and show that the social motives involved in religion are deep-rooted and essential. They are no other than those by which the individual self-consciousness itself is built up. To say, with Clifford and Romanes, that God is a "world-eject," a socialised and idealised "other-self," is to say that God is recognised naturally and inevitably, in the same sense that our human fellows are, on the one hand, and society as such, on the other hand. The differentiation of the personal object into myself and other-self, ego and alter, takes form gradually in a larger whole of personal values which are social in origin. It is this body of self-values, collective, ideal, mystical, that is ejectively embodied in God.

30. If this be true, religion will persist in human life and the religious interest will receive an interpretation that recognises these motives of man. Accord-

[1] The gradual secularisation of social institutions is of course a very notable fact. The "divine right of kings," the "temporal power" of the church, the "establishment" in all its forms—these give place to the radical separation of church and state. Similarly, there is the rapid secularisation of education, of law, of moral consciousness, with the growth of political liberty. All this means, however, I think, not the decay of the religious sanction, but the shifting of religious authority from a political and social to a spiritual and personal source. A similar development has taken place in the passing of the civil sanction as such from theocracy to democracy. There is a freeing both of religion and of the state, through the freeing of the individual. The growth of the individual in autonomy of judgment, producing a true democracy of conscience, refines the religious interest, but does not necessarily lessen or impair it. A greater menace is to be seen, perhaps, in the decay of the active religious practices in which the social motives of communion, revival, and common enthusiasm confirm and fortify the individual.

ingly, another interesting theory springs up, in which society itself—the original "other" and persistent fellow, to the individual—is considered the true religious object. According to Durkheim, to whom the evidence of the social origins of religion owes so much, the social group, whether clan, tribe, or nation, is the proper object, as it is the original cause, of the individual's religious interest. This interest forms for itself a personal God because it is unaware, or only vaguely aware, of its social obligations. What it really means to recognise and worship is the self, the spirit, the immanent principle, of Society.

We have here, in fact, a reasoned revival of a sort of religion of humanity—an interesting return to the line of tradition of French Positivism. It is sociological in spirit, a Comtean conception. Society "saves its face," as it were, in the presence of the individual in whom it has generated the religious need, by posing as God under certain thin disguises.[1]

31. The present writer signalised, in an earlier publication, the fact of the close relation between the god of a social group and its own national spirit, in terms which he ventures to utilise in the following paragraphs.[2]

"It appears in this way: the ideal self or deity to the individual, is the further carrying out, in the imagination, of the self-meaning; and this includes other individuals as well as the personal self. It is the ideal of a group, of a set of social relations, showing practical and moral oppositions, embarrassments and achievements. It is not the ideal held by other tribes and

[1] See Durkheim, *La Vie religieuse*, pp. 611 ff.

[2] *Proc. Fourth Inter. Cong. Hist. of Religion*, Oxford, August, 1908, reprinted in *Darwin and the Humanities*, 1st ed., 1909, pp. 101 ff.

races. The deity shows the growth of the normal social relations, and reflects their character, because he is the projected personal ideal of the group. While the deity must be thought of by these individuals as apart from them, since he is personal, yet he is the controlling spiritual presence, the voice, the oracle of the group, and may be approached through the proper mediation with rites and ceremonies. The tribal deity is in this important sense, then, the tribal spirit; he is conceived in terms of the tribal self.[1] The ideal that hovers over the personal self of the individual and impregnates his spiritual life, is one with that of the tribal or national self-consciousness. 'Great is Diana of the Ephesians' is not only a formula of personal religious experience; it is also a proclamation of civic or national unity; and both are possible in one because, in the process by which the individual idealises his life in community with others, he also erects, in common with them, a communal or national ideal. 'The perfect self,' he might say, 'which I should attain, is the same as that which you also find you should attain; and it is the same that we both imagine as our national spirit, patron, or God.' Deity may always be taken, therefore, to reveal the communal ideal of personality, as that develops continuously, while, at the same time, it supplies the appropriate object for the individual's personal worship. The Jehovah of the Hebrews is the embodiment both of the national aspirations, as voiced

[1] *Cf.* Espinas, *Les Origines de la Technologie*, 1897, p. 34 ff., who says: "The God of a people is nothing else than its own moral consciousness objectified. Zeus represents what is common to the ideals of the Greeks scattered from the Euxine to the Pillars of Hercules. Later on, when reflection became possible, Heracleitus seemed to understand this. 'The common reason,' said he, 'which is the divine reason, and through which we become rational, is the measure of truth.'"

in the religion of the prophets, and of the ethical qualities of the Jews. What a contrast to the polytheism of the Greeks!"

But it does not follow from this that the group, *as it exists*, is the object of religion. The existing group, the sociological group, is not what the religious ideal denotes, nor what the national aspirations celebrate. I do not sing to America *as it is*, when I sing the hymn "America;" but to America *as it should be*, its ideals fulfilled, its promise accomplished.

> "My country, 'tis of thee,
> *Thou land of liberty*—
> Of thee I sing!
> Land where our fathers died,
> Land of the Pilgrims' pride,
> From every mountain side
> *Let freedom ring!*"

This is a hymn to liberty, to freedom, of which America is taken to be the symbol. The social ideal is symbolised by the group, just as the ideal self or god of the individual is symbolised in the concrete object of religious veneration. The two ideals are fused in one through the motives under which—as we have seen—personality is at once socialised and idealised. But under this common ideal, imposed equally upon both, the concrete individual and the concrete group live in all sorts of opposition and disharmony.

Religion of humanity, then, to be a religion, must mean religion of ideal humanity; but this is what religion of divinity also means. For divinity is humanity idealised in both its aspects, individual and social.[1]

[1] Hence to find ground for identifying the religious object with society, we must seize upon illustrations of high patriotism, in which ideals are

32. The reason of the fusion of the two ideals, social and religious, is to be found in part, as has been remarked, in the factors by which they are in common established: factors intrinsic to the development of personality. But that is not sufficient. The outcome of idealisation in the social realm is not a supreme personality, but an ideal group, a utopian social order, for which all the individuals must be equally fitted.[1] This we have seen to be ethical in its character, not religious. Actually, it is realised only in part, and progressively, through the reflection in the group of the moralisation of its members. How then does this ideal come to merge itself in that of the religious life?

It would seem to be in order, as we may surmise, speaking teleologically, to find for itself a concrete embodiment.[2] The utopian society is conceived in

---

embodied: the flag, the national hero, the incident of patriotic virtue, the holy war, the crusade, etc. These do invoke ideals that seem to merge themselves in that of religion. But when we come to look into the common prosaic life of society—its *de facto*, un-ideal, "seamy" side —how the religious ideal finds itself repelled! Tradition, public taste, art, justice, institutions of all sorts, in which the *ensemble* of the group life shows itself, may be un-ideal and non-religious—often worse.

[1] It is an interesting question whether, in an ideal society, each and all of the members would have to be ideal individuals.

[2] This would seem to be in somewhat marked opposition to M. Durkheim's view (*loc. cit.*, pp. 600 ff.), who admits that it is ideal society that is symbolised in religion, but goes on to maintain that social ideals are in some way actually present in the social reality and are absorbed by the individual on occasions of "social effervescence." The "something more" presented to the individual on these occasions ("more," that is, than the usual social *rapport*) is the same as the "something more" by which, to the individual, the social ideal surpasses the social fact. This is an extraordinarily summary way of disposing of ideals! What are ideals, but imagined or desired ends set up in some consciousness? How can they be present in society before any individual conceives or imagines them?—unless, indeed, society be supposed to have an actual aggregate consciousness of its own. The ideal present in a society can

terms of its perfected unit. The formal obligation of morality does not suffice for the ends of ideal society, any more than for those of individual life. The postulate of the ideal group, even more than that of ideal individuality, must be reinforced by the assumption of the existence of a being who embodies them both at once.

In this sense again religion serves to bind together the actual and the ideal. Just as, for the individual, the reality of God substitutes a personal relation for the mere formal postulate of the ideal; so also for society it substitutes, for a utopian moral order, a genuine concrete end. The injunction, "Be ye perfect, even as your Father in heaven is perfect," means more, both to the individual and to society, than if it read, "Be ye perfect as a moral individual and in your social relationships." The element of justice in the social ideal, for example, has progressed through concrete historical stages, and is taking on specific international

---

only be the principle which seems to some thinker to explain its character and tendencies; but, so far as it is ideal, *it is not yet realised in society.*

Further, is it psychologically true that we conceive our ideals when we are lashed into social frenzy, carried away by social "effervescence"? If so, then the riot and the mob would present the occasion for the birth of social and religious ideals! This is to my mind nothing short of an affront to that fine and noble movement of consciousness by which it interprets its data forward to their fullest and richest meaning. It is irrelevant to deny, as M. Durkheim does, a mystic faculty of idealisation, for it is not at all mystical, nor is it a special faculty. It is, on the contrary, the normal counterpart to the judgment of fact; it is the assumption of value, out of which, in the form of confirmed hypotheses, new judgments are established and selected. This assumptive, idealising movement of the imagination is just as normal as the judging, believing movement is; and it is as essential to the development of knowledge. See the treatment of the entire subject by Meinong, *Über Annahmen*, 2nd ed., and the writer's *Thought and Things*, vol. i., chap. viii., § 6 f., and chap. x., § 8; also vol. iii., chap. viii., § 3.

form to-day in law, arbitration, treaties, etc., through the progress in the ideals of individual right; but these, in turn, have a further sanction and represent a stronger motive when, leaving the domain of pure ethics, they enter into the sphere of religion. "Thou shalt love the Lord thy God with all thy heart, and thy neighbour as thyself," is the counsel of justice, equality, and right of the religious ideal.

33. We are obliged to conclude then, that, instead of disappearing, religion in some form will abide. In the higher union of the motives of personal and social interest in that of religion, there is a return on another plane to the early state of things noted by sociologists in primitive culture, where religion dominated both individual and social life. The sacred symbol was the emblem of clan and tribe and also the badge of individuality and personal relationship. It is interesting to see that, after the successive differentiations of motive in the special interests of civilised life and reflective thought, the resulting ideals fuse together again in the postulate of Deity.

How far this postulate makes itself good, however, either in practice or for reflection, is another question. It requires detailed study. It is the philosophical question as to whether and how far the religious consciousness is really the organ of the apprehension of reality in the form of God.[1]

[1] Enough has been said, however, to show that primitive religion was not, as is often claimed, a first attempt at a logical explanation of the world. It was, on the contrary, an emotional and social interpretation, from which the logical and scientific points of view only gradually freed themselves.

CHAPTER VII

RELIGIOUS REALITY AND RELIGIOUS NEGATION

§ *1.* *The Religious Object as existing*

WHEN we examine more closely the factor of idealisation in the organisation of the religious interest, certain peculiarities appear.

1. As we have seen, the affective ideal is a term of value set up at the limit of the process through which ends are mediated by facts. The ideal man, for example, is the supposed final man, as he would be if the series of better and better actual men were carried on to infinity, to the ideal. The actual men mediate progressively the ideal or perfect man. At the limit, however, the means, the real cases, disappear; the ideal is allowed to stand in its own right. The postulation or assumption of an ideal ethical value does not rest upon hypothetical acts of virtue, but imposes itself as unconditional and absolute. This means, of course, that it is really a postulated value, not an actual one. So far as an analogous value is actually realised, it is part of the full reality of the thing to which its worth is attributed. As ideal, the most we can say of it is that it is possible and desirable.[1]

[1] This is the origin of the resulting norms of the "practical reason"; see *Thought and Things*, vol. iii., chap. vii.

In the movement of cognition proper while the result is similar, the process is the opposite; the idea or truth set up as means is isolated from the value it mediates, and is itself taken as a naked and neutral datum of reason. It is supposed to hold irrespective of the interest that apprehends it. The cognitive ideal is disinterested, personally neutral.[1]

In the case of religion both these processes seem to be modified.

2. In the first place, the ideal itself of the religious interest—the ideal personality of God—is set up not as simply ideal, simply desirable and possible, but as actually existing. The objective thing or idea which mediates the value does not fall away, as in other cases, in the process by which the ideal is postulated. On the contrary, it remains; and the ideal takes on the form of an actually realised worth. "God exists and has the ideal attributes; with reference to him they are not ideal, but real; it is only with reference to us that they are ideals": this is the utterance of the religious consciousness. There seems here to be an attempt to escape from the dualism characteristic of the practical life, thc dualism between fact or idea and end; to find in one experience both the real person and the absolute good.[2]

3. Moreover, second, the integrity of the logical movement is also impaired. The religious object, although taken to exist, is not the outcome of processes of knowledge of a logical sort—generalisation, impli-

[1] So the categories of the pure reason: see *ibid.*, vol. ii., chap. vii., § 4, and vol. iii., chap. ii., § 5.

[2] This has its counterpart in the history of reflection, in the attempts to find in God both the *summum bonum* and the infinite idea (in the terms of the first great exponent of this doctrine, Plato).

cation, reasoning. God is not a logical general or universal, following by necessary implication from concrete data; not a concept at all, but an imaginative postulate or assumption. We do not find here, in the contents of the religious meaning, a system of relational terms treated as the neutral and disinterested discovery of thought. Far from it; God is a personal presence, in the singular number,[1] which imposes directly upon the believer the obligation of respect and the duty to practise the religious rites. Its sanctions are those of an emotional and social order, rather than those of a physical or intellectual order. God is variously described by the faithful as a need, a resort, an inspiration, a saviour, a refuge and help; not primarily as a truth or a conclusion. And the attempts to prove the existence of God logically have never had more than the degree of cogency which attaches to apologetic modes of argumentation.

So far then from finding that the intellectual scaffolding gives body to the religious ideal, we find, on the contrary, that this scaffolding can not stand alone. The personal God is not even, like the personal fellow-man, an object of direct contact through the ordinary avenues of inter-subjective intercourse. In religion in general the physical manifestations of the presence of the deity, taking on the form of miracles, are food for the sense of wonder, awe, and mystery; but they are not the means by which he is normally apprehended or approached. And with the progress of logical modes of thought, the miraculous factor in the religious experience is greatly reduced. Along with this disappears also the claim of religion to explain the uni-

[1] Even in polytheism, the different gods are unlike one another; each has his appropriate province or domain.

verse by a special cosmology and a special psychology; to discover the laws of things and the rules of life by a special system of magical formulas or by a series of special revelations.[1]

4. Why then, we may ask, does the religious ideal, unlike other non-intellectual ideal values, impose itself upon consciousness as something actually existing?

It can not be because it is an ideal of personality; for the ethical ideal, which does not make this claim, is also such: nor because it is an ideal of truth; for it is not established by the processes of knowledge which discover and confirm the true.

It is, in our view, *because it remains always a social ideal*, an ideal of actual intercourse. Unlike the ethical and the intellectual, it never passes into the phase of inner autonomy to the individual to which, in both the other cases, we have applied the term "synnomic." Religion, considered both as a personal interest and as an objective content, remains "syndoxic" or aggregate in its force; it is a body of individual acknowledgments, of personal consents; it does not pass over into the realm of impersonal imperatives or norms. This distinguishes it both from the ethical and from the rational.

5. The contrast between religion and morals in this respect is especially instructive. The ethical passes from the form of mutual obligation, as between man and man, into duty to God as a fellow-person on a different plane, and finally becomes simply and only duty to the moral law, to the moral ideal itself, to one's own inner light, and to one's self.

[1] In Christianity the range of revelation and inspiration has been gradually restricted to spiritual things; even the infallibility of religious authority does not extend to secular matters.

"To thine own self be true,
And thou hast done with fears;
Man knows no other law,
Search he a thousand years."
SWINBURNE[1]

The moral law is thus self-imposed and self-sanctioned; just as the ideal of truth is self-consistent and self-sustaining. Neither requires a further existence, an external person, to maintain it. Merely by its presence in the consciousness of the one individual, it shows itself to be universal and necessary, "synnomic," in its force.

But the religious ideal is not of this character. It is not and can not be universalised in judgments of synnomic force. In its essence, as a first-hand experience, it requires the actual presence of the Other: God must be really at hand. His presence to one will not do for another; nor can the intimate character of the relation be generalised. In religion, one does not feel the force of a self-legislating and formal imperative.

§ 2. *The union of Ideal and Actual in the Religious Reality*

6. Looked upon as a way of interpreting the world, religion is historically of the first importance. It is a first interpretation, a form of interest focused directly upon certain things—things sacred, holy, divine. In this spontaneous form, as a direct interest, it is not to be confused with the theory of religion, nor with the philosophy[2] that justifies the religious in preference to other interpretations.

[1] Quoted from memory.

[2] This latter is to be taken up later on. The science of objective religions, it is plain, can not exhaust the sphere of religion, just as the "sci-

If we are right in finding the religious object, the deity, to be in some sense personal, as well as ideal, then we may say that the reality upon which it terminates is of the nature of a postulate; something supposed, imagined, anticipated, not something directly given, experienced, or proved.[1] Not that personality in others can not be actually experienced and in its own way proved, for purposes of knowledge—it can be and is: but that ideal personality can not be. All ideals are in their nature products of the imagination, built up upon knowledge, but going beyond it. This is true of the ideal person, God. Like the ethical and æsthetic, the religious ideal is a postulate.

The religious postulate, however, as racial history shows, preceded these others genetically. It was present when morals were merely socially sanctioned habits and customs, in which the properly ethical imperative, the inner light of conscience, had not yet appeared. The authority whose agents and penalties held society together in an effective organisation was in the first instance religious. It is probable that the first interpretation of the world, the first recognition

ence of manners" can not take the place of ethics; for in each case there is the personal attitude of the agent which is not fully embodied in the external facts or institutions studied. In the "science of manners" only the general forms of conduct can be observed, the mere shell, not the inner imperative or ideal with reference to which the individual makes his moral decisions. So with religion; institutions and cults, historically and comparatively considered, conserve values which must in each case be interpreted by the individual in terms of personal communion or absorption. But the two subjective experiences remain on different levels, since while ethical judgment has the synnomic or legislative force, religious faith has not.

[1] On this distinction, looked at genetically, see *Thought and Things*, vol. iii., "Interest and Art," chap. i., and vol. ii., "Experimental Logic" chap. 5.

of a force in and back of appearances in nature, was that of religion. In this sense, as Comte declared, the theological age preceded the metaphysical and the positive ages.

In all the differentiations that follow, in both racial and individual development, the religious interest retains its proper character. The practical and theoretical interests develop from this common root; but the main stem remains erect, producing another tree—the tree of religion itself.

7. The constant peculiarity of this interest is that it both idealises its object with the moral, and actualises it with the intellectual.[1] In this, as we have seen, it seems to partake of both the ideals of the practical and the rational life, different as they are; for practice idealises in erecting ends, and knowledge actualises in discovering truth. So we are driven to the question that this state of things suggests: Is it true that in religion we have a mode of experience in which the opposition between the actual and the ideal is overcome? Do the theoretical and practical interests of mankind come to a real reconciliation and union in the religious postulate of God?

8. We have to answer "no" to this question for two essential reasons.

In the first place, no postulate as such can of itself permanently satisfy the theoretical interest, the interest of proof. The absence of proof will always torture and the unfulfilled categories of reason will always protest. Witness the logical controversies as to the existence and nature of God; and the philosophical

[1] See the writer's *Darwin and the Humanities*, 2nd ed., London, p. 105. In that work (chap. v.) the social character of religion, as proved by anthropology and psychology alike, is put in evidence.

attempts to construe the categories of cause, identity, sufficient reason, law, in terms of design, personality, providence, freedom. The mere assumption of divine personality, whose rôle it is to accomplish this feat, in no wise shows how it can be logically done.

Whatever may be the value of religion as supplementary to reason, as substitute for it, or even as weapon of its refutation, it can not be said to fulfil the ideal of reason, which desiderates a body of neutral truth, independent of the personal life and interests of the individual.

9. A second and more positive objection is to be discovered in the nature of the religious experience itself. This experience does not reveal an ordered organisation of the factors of our varied interests, but a loose mixture of motives, each forcing itself on occasion to the fore.[1] It is in religious emotion, intense, ecstatic, overwhelming, that the actuality of the deity, the very presence of God, is vividly realised; but it is just in this experience that the ideality of the divine nature, both rational and practical, seems largely to disappear. God, the holy, the awful, the ineffable, the infinite, is, by reason of his very ideality, unapproachable, removed from human life and apprehension. As ideal, he can be contemplated only afar off; it is only as actual that he can be humanised. Hence it is that a mediator, a priest, an intercessor, a saviour, a madonna, a saint intervenes to stand, as an actual presence, between God, the ineffable ideal, and the finite man. This movement—common to different religions—and substitution of a more-human person for

[1] The conflicts and oppositions of the religious life have been well brought out by Stratton (G. M. Stratton, *The Psychology of the Religious Life*, 1911.)

God or of a more-holy person for man—is sufficient witness to the failure of the religious interest to reach of itself an object at once actual and ideal.[1]

What would seem to be lacking is a religious "imperative," an intuition of religious reason, imposing belief in the existence of a personal God. But just in passing into this form—if religion did pass into it—thus becoming synnomic, as thought and conscience are, it would lose the social character that gives it concreteness and actuality. God would then be, like rational identity and moral law, an empty form, the postulate of value in an abstract realm; a norm of reason, instead of the actual person found by the devotee to be a "very present help in trouble." The warmth and intimacy of personal religion would be lost to humanity, much as the child misses the sympathetic counsels of a loving parent, when, on arriving at maturity, he finds it necessary to be a moral lawgiver to himself.[2] The divine attributes of pity, mercy, pardon, love, are personal and in a sense *optional* with the deity. In the lower religions, they are represented in terms similar to human caprice or self-interest; the deities have moods, are subject to fatigue, ennui, and passion. Refine these personal attributes

[1] The "mysteries of faith," many of them common to different religions, are properly called mysteries: the union of divine and human personality, the incarnation, the virgin birth, the bodily resurrection. They all assert for faith the presence of the infinite and ideal in the supernatural; but what they present to the eyes is the finite and actual. Beyond faith, there is no further assurance.

[2] In the matter of the religious "instinct" leading men, when in danger and in crises of every kind, to call upon God, I think we are in the presence, not of a form of religious imperative proper, but of a form of social dependence in which the elements of the religious life are in evidence. It is like the appeal of a prisoner to the judge for clemency or mercy; an appeal outside the domain of justice or reason.

as we may, they can not be universalised in formal statements.

10. The actuality of the religious object, indeed, is not guaranteed by knowledge or proved by reason; it must be accepted by faith. It is a matter of personal experience, social both in origin and in nature. The social relation of man to God, as one of dependence, help, communion, is written large on every page of religious narrative. The dualism of self-and-other is sharpened as the individuality of the believer is heightened over against that of God; and when, in religious ecstasy, the individual loses himself by participation in the mystic presence, the religious interest properly speaking loses itself in turn in the general fusion of the elements of personality.

§3. *The Religious Antinomy*

11. But if the element of religious actuality is immediate, that of religious ideality is mediate. An ideal is always the end-term of a process of mediation, whether this be theoretical or practical. Ideal truth is a supposed extension of actual truth; it is an outcome, mediated by ideas, feelings, experiences that are actual. Likewise ideal value is something beyond the present actual worth we possess. The latter as means mediate the former as end.

Accordingly, it is plain that if the ideality of God is insisted upon, it must be by reasserting the finiteness and incompleteness of the means—the symbols and emblems of the religious life—by which this ideality is mediated. Hence the force and success of the reformations in history which protest against the tendency to "idolatry" and "image-worship," that is, the tendency to identify God with the image or idol that symbolises his presence. The purity of the ideal is in constant danger of being impaired by the return to actuality as represented in the symbol.

But the reverse tendency is also real; and it issues in a movement that is equally in evidence in the history of religion: the protest of the practical religious life against the abstractness of creeds and the vagueness of philosophical definitions. In Deism and Pantheism the religious interest finds much to complain of—the remoteness and impersonality of the Deity. It also finds artificial the intrusion of a mediator or intercessor, brought in to guarantee and actualise a relation which should be one of direct experience and communion.

12. These two historical tendencies taken together show the essential factors of the religious ideal—its actuality protesting against its ideality, and its ideality explaining away its actuality. They illustrate, better than any theoretical statement could, the inherent difficulty of maintaining the actuality of an ideal erected as a postulate by the processes of affective logic and interest. It also illustrates the difficulty of reading any concrete experience in terms of completed worth or ideality. Ideals are not reached as discoveries, they are what we desire; actualities are what we discover, whether we find them desirable or not.

This sharp antinomy in the meaning of the religious object makes the reality it postulates unsatisfying as a synthetic mode of experience. The dualism under which consciousness labours, in its various renderings of the real, are not relieved; the embarrassments of the practical life are not removed. The resort to the supernatural is the inevitable outcome of the development of the religious interest; but on the terms of this postulate, the supernatural enters into the natural only as a *deus ex machina*, by a miraculous intervention. It is a supposition motived by fundamental demands of the moral and religious life; but it is not established as part of the system of existing realities. The oppositions between ideal and actual, the postulate and the implication, the assumption and the presupposition, the mediate and the immediate, are mitigated, it is true, but they are not radically healed.

§ 4. *Religious Negation. (1) The Non-Religious or Secular*

In the consideration of the various modes of psychic function, it is profitable to investigate the negative along with the positive phase of meaning.[1] We may well carry further the enquiry as to the existence of negation in the movement of religious interest.[2]

13. A somewhat radical distinction has been established, it will be remembered, between the attitudes of denial and rejection.[3] Denial attaches—along with affirmation—to a proposed relationship in the content of knowledge or thought; this relationship is denied when it fails to establish itself in judgment. One thing, the predicate, is denied of another, the subject. Logical denial is an advanced stage of this sort of opposition between items of knowledge.

Rejection—on the contrary—along with acceptance, is a different attitude. It is the attitude of refusing, repelling, withdrawing; of excluding something from

[1] Cf. *Thought and Things*, where pre-logical, logical, affective, and æsthetic rejection and denial are considered.

[2] It is considered, for the case of early racial interpretation, above, chap. iv., sects. 8–10.

[3] See above, chap. iv., sect. 8 f.

the sphere of control or existence in which it might have found a place. It is a movement of interest and selection, not one of mere recognition or acknowledgment of fact or truth. Denial is merely recognitive of an exclusion which holds between terms; rejection is selective and personal, a matter of the exclusive interest of the knower himself. Denial is a matter of inability to believe; rejection, of the refusal to tolerate.[1]

But in negation by selection, there are again two cases, as we have seen above. In selecting something as good, beautiful, satisfying, we may simply overlook, forget, or ignore, and in this way exclude, everything except that to which our interest is directed. When our exclusive interest selects something, all the rest of the world is for the time rejected. Æsthetic enjoyment is notably of this sort—absorbing and exclusive. This has been described as "privative" exclusion or negation.[2]

But there is a more positive rejection, an intentional exclusion, that means more than simple neglect. The morally bad, the æsthetically ugly, are the objects of actual rejection, of positive condemnation and disapproval.

In these two forms, both of which are characteristic of the life of interest, selective exclusion is in striking contrast with cognitive opposition and logical denial.

14. We find, even on slight examination, that in the manner of its rejection, religious interest is true to its nature; it establishes both forms of selective exclusion, but it does not issue in logical denial.

In the case of primitive religions, the facts are plain.

[1] See *Thought and Things*, vol. i., chap. ix., and vol. ii., chap. viii.

[2] *Ibid.*, vol. i., chap. ix., § 4; also vol. iii., chap. ix., and chap. x., §§ 5 ff.

In the totemic system, the animal totem includes the group; this in turn excludes the members of other totemic groups. It is not by reason of different individual characters or qualities that this exclusion takes place; it is by reason of the totemic name and the relationship signified by it. The affective interest by which the social and mystic class is established rules out all but members of this class, no matter what their other qualifications may be. This is privative exclusion; it proceeds by the establishment of a class through a movement of interest; all else is for this interest negative.

Within the same phratries (the larger divisions of a tribe) various clans may exist, with their totemic symbols side by side. In certain of their interests and practices—such as intermarriage between the clans—these groups overlap and interpenetrate; in others—such as the eating of the flesh of the totemic animals—they are rigidly exclusive. But in neither of these cases is the exclusion of the nature of a logical opposition based on generalisation or induction.[1] It is not due to cognition of differences, but to the selections and exclusions of interest.

This state of things extends to family relationships, tribal organisation in time and space, in short to all the details in which the totem stands for the group. The other class, that of the foreign totem, is in so far not considered profane, not necessarily hostile, not at all wrong in its similar adhesion to its totem; its affairs are merely neglected, not included in the interest of the one who finds it different from his own. From the

[1] M. Durkheim appears to confuse the two, or at least not to distinguish them, in saying (*La Vie religieuse*, p. 342) that the savage's negations are simply "excessive" logical denials.

point of view of the sacred, it is secular. But it has its own system of sacred things; each allows this to all. Intolerance as such does not begin in the sphere of the secular, but only when we touch upon the more positive sort of exclusion characteristic of the "profane."

The secular or non-religious in general is thus defined: it is all of that which is outside of the religious interest as neither belonging to it nor opposed to it. It is the religiously indifferent. That, on the contrary, which is opposed to the religious interest is not secular, since it is not indifferent; it is the profane [1]

§5. *Religious Negation.* (2) *The Irreligious or Profane*

15. Over and beyond the mere indifference of the religious interest, which defines the secular, there is a more positive movement of exclusion or rejection, which defines the profane, the irreligious, the religious bad. All through the life of the primitive clan—to cite primitive religion again—there runs a system of positive prohibitions. Innumerable things—marriage within the clan, intrusion upon sacred places, eating of the flesh of sacred animals, contact with things which are taboo—are tragic in their negative meaning. The religious bad like the ethical, later on, is something to avoid, to resist, to combat, to destroy. The social motives organised in the religious interest reject the profane thing, and in so doing, define it.

All religions alike show this: all have their profane, as well as their sacred; and none stop with mere logical definition of the one or the other. Religion gene-

[1] I do not find anywhere this or indeed any clear distinction between the merely secular and the profane; discussions generally treat them together, as opposed to the sacred.

rally personifies the evil or profane in demons, evil spirits, devils, thus rendering it more positive and concrete. As the sacred, so too the profane is made incarnate. Satan is the arch-enemy of God, the source of all profanations, the "great-bad," to be rejected in all his forms by society and by the single man.

In another place[1] we have discussed the topic of the supremely or "ideally" bad, the *summum malum*, in treating of the ethical mode of negation. We saw that the process of idealisation, following its normal course, imagines the fully evil, the great-bad. It is the "ideal to avoid," something set up by way of contrast to the ideal good, but not found existing anywhere, except in the religious postulate of a supreme devil or Satan. The assumption of a moral ideal, imposed by the practical reason, does not involve the reality of a negative ideal, the supremely bad, but only defines it in opposition to the ideal good.

16. In religion, however, as we have seen, the postulation of the ideal carries the assumption of actual existence: God exists. And in this assumption, actuality is given also to the ethical demand for the good person and for a perfect moral or social order. God is a person, and with him exists a "kingdom of heaven," peopled by saints and holy spirits. It is interesting now to note that the motives of religious exclusion in the definition of the profane involve the same demand for actuality. Not only does the devil, the infernal person, exist, but there is also a kingdom of evil, a place full of spirits, personal but malignant, who do violence in every possible way to the ideals of religion and morality. The hierarchy of saints, angels, seraphs, archangels, is matched by that of demons,

[1] *Thought and Things*, vol. iii., chap. ix.

devils, arch-devils, Satan. All these exist by much the same right of actuality as the good spirits and God—the right of the social fellow whose presence is established with that of the person to whom he is the *socius*.

While this is undoubtedly true—the history of religion fully proves it—there are still reasons for thinking that in this wide demand for actuality, religion in its cruder stages asserts a claim which is later to be abandoned or is at least not insisted upon. Granted that in the social and concrete character of the ideal the actual existence of God is involved, does it follow that this is true of lesser personalities, either good or bad?—and is it true of the supremely bad personality, the arch-devil?

§ 6. *Profane Reality: the Arch-Devil*

We have to answer this question negatively in both cases. In the subordinate personages of all kinds, the ideality is lacking which the religious interest desiderates; they are carried along with God by a *diffusion of sacredness*, which is a most common phenomenon in the sphere of religion. In the arch-devil, on the other hand, the social requirement of religion is violated: the devil is anti-social, as he is anti-moral.[1]

17. As to angels and demons alike—creatures that come short of being gods—their very incompleteness and subordination render them unavailable as independent ends of genuine religious interest. As soon

[1] Apart from considerations of origin, we might say that the essential religious relation between the one person and God might subsist regardless of the existence of any other persons whatever in the universe; and also, on the contrary, that, while not demanded, personalities short of gods, like angels and demons, may exist by the same right—whatever it be—that good and bad persons exist in the world.

as the postulate of one supreme deity, the ideal self of the refined religious consciousness, disentangles itself from the social conditions of its origin, this ideal stands alone. It passes from the cruder stages, in which it earlier embodies itself, into the region of spiritual perfection. The subordinate beings remain as vestiges, a kind of reduced gods, or magnified men—epic mythical heroes or ancestors, beings having all sorts of semi-divine and semi-satanic attributes. They constitute a sort of filling in of the series of mediating terms between man and God, a series which the religious interest continually seeks to constitute in order to bridge the interval between the finite and the infinite, the actual and the ideal. But this is not essential to the postulate of God. In the development of Christian theology, for example, the theory of angels has always been a sort of romantic annex to serious doctrine, a concession at once to a stronger faith and to a mysticism tinged with the æsthetic[1]; and demons have been the gargoyles, as it were, on the temple of sacred truth. Neither the one nor the other is taken too seriously when once religion has outgrown the swaddling clothes of mystic participation and myth, and has substituted the truths of psychology and science for the practices of psychosophy and magic.

18. The case of the devil—that is, the arch-fiend, as opposed to mere demons—is more instructive, both because of the semblance of actuality thrown about him by religion, and because of the fact that reflective and critical religion itself understands that his existence in personal form is only semblant. As has been said, the demand for concreteness and social realness in the relation of God to man leads to the assumption of

[1] How angels have figured in sacred art!

actuality; and this assumption passes, by the same process and for the same reason, over to the great-bad, the devil. As one can worship a living God more fervently, so one can fight more effectively a living devil. So the "great-bad" of morals becomes the "infernal person" of religion.

But both of these, the great-bad and the infernal, represent in the imagination not ideals of acceptance, of the desirable, but ideals of rejection, of the undesirable. They are things to avoid, to spurn. They are inconsistent with the reality of the great-good, the ideal worth, and the divine person. The *summum malum* is in opposition to the *summum bonum*. Even in ethics, where the postulate of actuality is not present, only one of these values, the good or the bad, can be considered supreme.

When transferred to the field of religion, this opposition becomes actual conflict. The world becomes the scene of mortal combat. The devil is the incarnation of the principle of disorder, evil, and sin—he is the anti-social and anti-ethical ideal. He brings his cohorts with him and establishes in the world an annex to the inferno, the kingdom of evil. But he is not a being equal to God, not a peer of Divinity, but a presence which has been many times routed and which will be finally crushed, as in the splendid apocalyptic vision of Saint John. The devil is a sort of cosmic opposition party which must be kept strong and effective, a thorn in the side of the good; it can never, however, be more than just-short of supreme.

We find, therefore, that the negative ideal, a fully malignant character, a devilish god, is not and can not be assumed to exist as a fact; it is only a simulacrum, a semblant being, victory over whom for good and all would destroy evil and bad values of every kind. But the bad values continue to exist, and the devil is assumed to impersonate them.

In other words, put technically, the existence of the ideal personality or God, destroys the pretended actuality of the other and opposed ideal, the bad-spirit, the personal devil.

19. It is only in a state of religious development akin to polytheism, or in an atmosphere of religious dogmatism that stifles reflection, that belief in an actually existing devil of this semi-divine kind is to be found.[1] The deity of the primitive tribe is one among many, each of which is represented as guarding the interests of his own tribe and being guarded in turn by it; he is represented as sacred for his people and supreme; although for others he is a false god, a profane belief, a rival. In more advanced culture, God is made a personality of so concrete and human a nature that his life is looked upon as one of struggle, of alternating victory and defeat, achievement for and with his people. The enemy, in this case, is the actual devil; not merely the god of another tribe, but an evil person having many of the attributes of divinity and capable of coping with the God himself, though at a disadvantage. In both these cases, the ideal has not

[1] We have noted above in passing the fact of the "diffusion of sacredness" over a wide range of things related to religion. It is a phenomenon of the transfer of interest when emotional and imaginative factors are dominant. In primitive life, it is very marked: everything at all connected with religion comes to reflect its values. Even with Christianity, we have the "sacred book" kissed in the law courts, biblical stories, even when immoral, read in open church services as "sacred lessons," chapters of profane and vulgar Jewish history incorporated in the "sacred text," and even the devil treated with marked respect as belonging in the region of sacred things. One must not swear by the name of the devil, as he must not by the name of God!

yet attained that exclusive statement which forbids the supposition of a second divine or semi-divine person, a rival to the deity.

20. The recurring concept of the devil serves to set forth most plainly the antinomy between the ideal and the actual inherent in the religious postulate as such. If the demand for actuality be taken literally, so literally as to carry the assumption of the existence of the devil, the ideal finds itself mutilated: the bad becomes actual fellow to the good. If, on the other hand, the ideality of God be consistently affirmed, the marks of actuality are one by one effaced. The essential postulate rests in the region of faith, by which the divine is accepted as both actual and ideal. But the contradiction is lessened and the antinomy made less embarrassing for reflection, if while the possibility of lesser spirits of both sorts be allowed, that of the arch-devil be disallowed. The semi-divine and the semi-devilish may exist; but the supremely bad, co-ordinate with the supremely good, can not exist.

§ 7. *The Philosophy of Religion: Religion as organ of Value*

21. It would seem, finally, that the philosophical views known variously as religious realism, moralism, and theism, so far as they recognise as valid the postulate of a supreme and ideal personal God, can not be pluralistic. If this postulate is to be honoured in our reflection, it must be on its own terms; namely, the existence of an ideal which asserts itself by *selective exclusion.* Other personalities good and bad, other things secular and profane, may exist in the same universe with God, but not other equal spiritual ideals.

This is the monistic basis upon which alone a spiritualistic philosophy of religion can be constructed.

On the other hand, to accept a religious pluralism or polytheism—any form of plurality of good and bad spirits contending or allied *inter se* on equal terms—is to go back to the earlier anthropomorphic stages in which the religious ideal itself involved a plurality of personal agencies, existing in relation to one another. This is the basis of a theory of pluralistic realism or moralism.

These two alternatives reflect respectively the terms of the essential antinomy of religion as spoken of above, ideality and actuality.

22. The meaning of religion for our further thought is now fairly plain. It is an interest in which the earliest collective and personal values of human life were organised, and in which also the later human values receive a synthetic and ideal statement.[1] God is the final and comprehensive value of the life of feeling and will; and as reality, this postulate gives concreteness to the ideal contained in the whole series of social and moral values. In this consists the dual claim of religion—not entirely made out as we have seen—to be the organ of the most profound apprehension of the nature of things. But apart from this claim we may say that it represents the type of reality found in *value*, carried on to infinity.

[1] Cf. Höffding, *Philosophy of Religion*.

CHAPTER VIII

LOGICAL INTERPRETATION

§ *1. The transition to Logic: the rôle of the Imagination*

THE transition in individual development from the prelogical to the logical, from the emotional to the theoretical interpretation, has already been touched upon. It is a movement in which the factors of emotional and collective interest gradually give way to those of thought. Exact observation, judgment, proof, come to a greater or lesser degree into their own.

1. We should recall here two points which are necessary to the understanding of the corresponding historical movement.

(1) In the passage from the prelogical to the logical type of knowledge, the imagination is the constant instrument. It establishes semblant combinations, schematic programmes, constructions of fancy, which present the tentative outlines from which the confirmed judgments of fact and truth are to be developed. By the schematising imagination, the materials of knowledge are released from the grasp of external and social control, and made available for reconstruction in experimental hypotheses and æsthetic unities.

In the function of play in childhood this imaginative semblance notably appears. The impulse to play shows itself in the exercise of untrammelled fancy.

The legend, the fairy tale, the dramatic incident of the school or the voyage, the detail of personal display or patriotic pride, all give occasion to a literal play of the imagination, from which the new freedom of thought is to emerge. The playful creation of fancy yields to the ready assumption, this to the sober hypothesis, and this in turn to the grounded proposal and the perfected theory. Thus the gains of knowledge become constantly the starting point of the imagination, until the ideals of truth and right and beauty are projected forward in the postulates of reason, duty, and God.

(2) In the outcome the great dualism of spheres, inner and outer, is established in which these imaginative programmes find themselves alternatively controlled, corrected, and confirmed. Mind and body, the spiritual and the physical world, take their place as opposing substances existing over against each other.

2. The same typical processes stand out in the movement of racial interpretation.[1]

(1) The outstanding characters, as we have seen, of primitive or prelogical racial thought appear in the prevalence of collective or emotional interest, and the dominance of mystical and religious motives in the interpretation of things generally. It is in these interests, in the bosom of social and religious concerns and observances, that the racial imagination stirs to produce safer and saner knowledge. In the social festivity, the dance, the game, having musical and ceremonial accompaniments, we find in savage life the outlet to emotion and caprice. In the license of the religious festival the bondage to social form, so compelling in all the details of serious life, is temporarily broken. The

[1] *Cf.* for detail the interpretation contained in the writer's *History of Psychology*, vols. i., ii.

most stringent regulations may be relaxed, the most serious lapses condoned[1]; even that which is customarily forbidden may be permitted or enjoined in the exceptional usages of the festival.

3. (2) Further, on these occasions, the imagination has a certain inventive rôle. The procedure follows a dramatic and serial order, as embodying a story or a series of cosmic or epic events. A mass of symbolic meaning is organised playfully, symbolically, or dramatically, just as is the case also in the playful dramatisation of civilised children's games.

In this there is a release of the motives of individualism. A greater freedom arises for the exercise of individual thought and imagination; an impulse asserts itself toward the explanation of things, in the whole range of nature and mind.

4. In religion, these movements take on permanent and semi-logical form. The religious life seeks symbolic embodiments, since it carries in it the beginnings of cosmology and theology. Hence the rise of myths and of mythical interpretations of nature in terms of religion: mythical cosmic events celebrating religious or national heroes, mythical accounts of supernatural situations by which the natural are explained. In the mythology of many religions, and in the sacred books of many others, we find the racial imagination making out its programmes, arranging its schemata, presenting its explanations, in the direction of a logical and reasonable understanding of the world. As in the individual, so here, the imagination entertains as plausible, probable, as-if-actual, what may turn out to be logically unreasonable or physically impossible. For in this its early rôle, the imaginative interpretation continues

[1] Such, for example, as those of sexual restraint.

to serve the ends of social tradition and religious dogma. The myth is apologetic of faith, even when explanatory of nature. The great events of the world are explained by beings and forces of which religion gives the true account. The promise of God after the flood explains the rainbow, as the resting of God on the seventh day explains the Jewish sabbath and with it the length of the seven-day week.[1]

5. This first resort to a mode of individual thought, imaginative and uncontrolled as it is, is in itself a notable achievement. Besides having instrumental value, as leading on to logical process, it throws the matter of representation into æsthetic and poetic wholes. Even when more sober scientific observation has superseded it, in the direction of a cosmology or a psychology, it still remains, maintaining itself as a realm of free fancy, of poetic creation. The myths of the seasons and of other recurring events of nature, of the rising and setting sun, of the movements of the heavenly bodies, of the origin of heaven and earth, of the music of the spheres, are epic poems in which the *dramatis personæ* are cosmic forces personated by gods and heroes. What is more beautiful than the story of the rainbow, that wondrous bow above the cloud, which records, in the presence of the sun, rising resplendent with all his joyous gifts of light and warmth, the divine pledge to dispel the mists and stay the flood! Or the

[1] In the successive volumes of Frazer's great work, *The Golden Bough*, 3rd ed., we have presented to us certain typical motives which are recurrent in religious myth and story (such as that of substitutive or vicarious sacrifice in Part vi, "The Scape-Goat"). These imaginative ways of interpreting social rules and religious usages in semi-logical form and in semblance of external fact, and of thus satisfying the curiosities of a too-prying faith, constitute a wonderful chapter in the true romance of thought.

myth of the six-day creation, according to which the Lord rested the seventh day, viewing the finished work of his hands, contemplating the perfection of that world in which his infinite wisdom and will were embodied! Greek mythology has served as a body of symbolic material from which the greatest artistic geniuses of all ages have chosen their noblest themes.

This latter, the intrinsic æsthetic value of the myth, has not had proper recognition. It is not only true that the imagination serves as an instrument to knowledge, by freeing the motives of individual thought and by prospecting for the objective and real; it is equally true that its constructions *may not have such ends in view*, but may constitute a mode of interpretation having independent meaning.

Here, in the race as in the individual, the resort is to a realm in which the mind escapes in some measure from the controls, social, factual, moral, of the serious and prosaic life, and feels itself allied with the mystic and romantic forces and beings of another world. The moral meaning of the rainbow is as high, and the lesson of its presence as inspiring, after the laws of scientific optics have come to replace the fiat of a personal creator. The interpretation due to poetic vision is reconstituted after every new discovery by which scientific truth is at once dislocated and rearticulated. The lengthening of the six creative days into geological periods, whatever scientific advantage it may have, does not impair the wonders of the creation as presented to the eye of contemplation. The starry heavens were the object of the same wonder to Immanuel Kant as to Pythagoras. The legends of the fall and redemption of man are semblant pictures, æsthetic figures, depicting inexpressible things, in the same sense as the Gothic cathedral and the solemnly worded Confession of Faith. Religion appeals to art, of which the myth is an early form, to present the divine-human drama to the eye. And this semblant picture, presented to faith and contemplation, is not merely a temporary substitute for a fully rational account; it is a permanent rendering of ideals in forms with which the logical dispenses, but which nevertheless hold their own in human thought.

6. The outcome on the side of theoretical interpretation is the dualism of mind and body, stated consciously as an opposition between the immaterial or spiritual and the material or physical. The doctrine of the soul in primitive thought did not realise this distinction fully. The notion of a conscious spiritual principle, the reason, different in nature from matter, did not arise in Greek philosophy; it became clear only in the writings of Saint Augustine. In Descartes it was for the first time made the presupposition of a logical interpretation of the world.

Allowing this brief sketch to stand as sufficient for our present purposes, we may now enquire as to the character of the logical interpretation which comes with the rise of conscious reflection.

§ 2. *The Problem of Reflection*

7. With the passing of the explicitly theological period of human interest, the religious point of view, which carried with it in early times the collective and emotional, yielded in this sphere and that to the secular and the intellectual. Industry, politics, science, art, each had to achieve a relative independence, as representing a vital interest of mankind. And while these

interests have never entirely superseded the religious in its own sphere, still they have succeeded in dividing the territory in such a way that conflicts are avoided, and problems are met by resort to distinctive and appropriate methods. Industrial methods no longer comprise sacrifice to the wind-gods, nor do political methods recognise theocracy and the right of divine interference. Likewise in the realm of knowledge and practice, a technology had to be worked out to replace that of religious rite and magical ceremonial.[1]

8. The Greek period was of great importance in this respect. It saw the gradual evolution of the problems of reflection, notably the problems arising about the relation of mind and body. And it served to teach the race with what instruments the mind might, in its own right, approach the world to understand and subjugate it. Prudence in practice, formal logic, resignation, moderation, scornful irony, subtle speculation, æsthetic intuition—all these, besides religious veneration and collective piety, were in turn the means by which the individual nourished his hope and renewed his courage in confronting the world.

In the result, we find modern philosophy starting out with a fairly definite programme. The alternatives of thought were given in the terms of the Cartesian dualism[2]; and those of method were anticipated,

[1] It is hard to overestimate the importance of method or technique everywhere, or the difficulty of destroying a false technique when it is entrenched in social and practical interests, and when there is no adequate procedure at hand to replace it. Rather than do nothing to avert disaster or to assure the favour of fortune, men will continue to do what they no longer believe in.

[2] In the little work cited (*History of Psychology*, vol. i., London ed., p. 95 ff.) the relation of Cartesian to later speculation is stated as follows: "The terms of the distinction between mind and body being

if not clearly formulated, in the directions of physical science, on the one hand, and philosophical speculation, on the other hand. Mind and body were the two realms in which science was to develop into the sciences; and these were also the sorts of existence or reality which philosophy was called upon to interpret.

In these terms the thought of the race formulated the opposition between the two substances or kinds of reality, much as the individual also formulates it, proceeding upon the two controls, inner and outer, found operative always in experience.

9. Further, it is clear that the movements of individual thought which issue in these two types of existence, suggest the methods which reflection employs to interpret them. The movement of knowledge proper is experimental, positive, confirmatory; by it the sphere of the actual is established and extended. This becomes the method of science, broadly defined. On the other hand, the movement by which practice develops, securing gratifications, defining values, erecting ideals, remains that of interest and sentiment, motived by utility and satisfaction. Both of these

---

now understood . . . speculation takes the form of an interpretation of this dualism itself. If we look upon the earlier thought as being a spontaneous or direct consideration of nature and man, we may look upon the latter as being a reflection upon the result of this former thinking. The dualism itself becomes a presupposition or datum; its terms condition the further problem. How can mind and matter both exist and give the appearance of interaction?—which of the two is the *prius* of the other?

"These questions as now formulated show later thought to be an *interpretation of dualism*, as the earlier was *an interpretation of the world issuing in dualism.* While the ancient and mediæval philosophies developed a progressive distinction and finally a divorce between body and mind, the modern results in a series of attempts to accommodate them to each other again in a single cosmic household."

are taken up in reflection and developed in theories which define reality respectively in terms of actuality, in the forms of fact, truth, rational coherence, and of ideality, in the forms of the good, the beautiful, God. The one type of theory turns upon truth, the other upon value.

§3. *Logical Theories: Scheme of Treatment*

10. In these several types of reflection the methods and results of spontaneous thought are taken up and justified. The actual is reached by the reflective recognition of the control, external or internal, in which spontaneous thought finds its trans-subjective reference. The spheres of the external and internal reference of ideas, the realms which this and that experience mediate, become the realities of the corresponding modes of reflection. The actual is thus manifold: physical, mental, conceptual, relational, and other. All these modes of actuality are confirmed by reflection, after being achieved by the processes of prelogical interpretation.

So, too, of reality of the ideal type; it has its reflective forms following upon the prelogical forms. The ideal is a further worth suggested and mediated by the present fact or idea. It becomes the reflective as before it was the distant and perhaps unconscious goal. There are, accordingly, ideals physical, mental, moral, social, religious. The reality reached by idealisation has many forms, as that reached by processes of actualisation also has.

11. In the history of reflection, great systems of thought have arisen based respectively on one or other of these alternatives. Theories which confine reality

to the actual are Materialism, realistic Spiritualism, Rationalism (in some of its forms), Positivism, Naturalism. On the other hand, the theories whose cornerstone reposes on ideality are Voluntarism, Theism, Pragmatism, Moralism, Æstheticism.[1]

12. Instead of recognising as valid the results and methods of spontaneous thought, however, reflection may criticise or in some way alter them by its interpretations. In fact, in recognising and justifying the two great modes of mediation—mediation of truth and mediation of value—reflection already renders a criticism, although in the first instance favourable. But such criticism may, on various grounds, issue in the denial of the validity of one or other or both of these mediations and attempt an adjustment of their results in a larger synthesis; or it may reject the presuppositions on which they proceed. In this last case, we find a more or less explicit resort to immediacy: a reflective justification of the sorts of apprehension in which consciousness seems to realise the existent without having resort to any process, either discursive or teleological, involving representation by ideas. The real may be that which most fully and directly realise, rather than that which we can prove or that which we desiderate as gratifying. Instead of the experience

[1] One would naturally use the term "idealism" for this entire group of theories, were it not for its ambiguity. Idealism is widely used for theories of the rationalist or intellectualist type in which reality is mediated through "ideas"; thus understood, it is equivalent to "idea"-ism. *Cf.* the note to § 10 of chap. x. In this sense, idealism falls in our group of actuality theories.

The ideality theories—the true "ideal"-isms—are generally less one-sided than the actuality theories, since it is harder to deny the actual in maintaining the ideal, than it is to deny the ideal in maintaining the actual. Being less synthetic and comprehensive, on the other hand, the actuality theories are truer to type.

in which ideas are actualised or that in which values are idealised, that which most fully reveals to us the inner nature of things is the experience in which the meaning of things is *directly realised.*

13. This is the motive of a further group of philosophical theories: a motive, however, not always clearly defined or unambiguous. The sorts of immediate experience resorted to differ in type and in value. We have before distinguished[1] three modes of immediacy, each having its peculiar place in mental development: the immediacy of "primitiveness" (as in sensation), that of fulfilment or "transcendence" (as in intuition), and that of "reconciliation" or synthesis (as in æsthetic contemplation). Each of these has been taken, in some historical theory, to be the most important or the only mode of apprehending the truly real.

Such theories may be classed together as "affectivistic" in opposition to those based on knowledge and will, which are called intellectual and voluntaristic. Among them we may cite Sensationalism, Intuitionism, Mysticism, Mystic Pantheism, Æsthetic Immediatism or Pancalism.

14. We find, in short, in the development of speculative theory, the recognition in turn of each of the great modes of mental function as being fundamentally determining, in the apprehension of reality. Knowledge, will, and feeling become, each in turn, the recognised organ of the true experience of the real. The actualising experience is one of knowledge, the idealising experience is one of will, the realising experience is one of feeling.

It would seem then that, historically considered, speculative thought has allowed, tacitly or avowedly, the presupposition that it is in a mode of experi-

[1] *Thought and Things*, vol. iii., "Interest and Art," chap. xiv.

ence, or *through consciousness*, that reality reveals itself. We may say that this is true without doubt. The subjective point of view, reached in the Socratic period and consciously adopted in the Cartesian, has remained the starting point of the theory of reality, as it is the presupposition of the judgment of existence. A theory may refuse to admit this and attempt to construe reality as something quite apart from consciousness or knowledge; but in that case it must still postulate a principle—that of matter, or that of mind, or that of God—the meaning of which can be determined only in human experience. Such a principle loses all determination and value for theory apart from experience. The "infinitely infinite," the "unknowable," the nirvanas, are not positive principles, but ontological postulates formed in advance to meet the exigencies of abstract thought. This remains, however, to be taken up again in our later discussions.

15. The foregoing division of theories may be presented to the eye in the following table, which will serve to guide us in the brief expositions and criticisms that follow.

TABLE III. THEORIES OF REALITY

A. MEDIATION THEORIES	I. Actualisation theories, finding reality in facts, truths, principles.
	II. Idealisation theories, finding reality in ends, values, norms.
B. IMMEDIACY THEORIES	III. Realisation theories, finding reality in direct fulfilments.[1]

[1] In lieu of a better, I employ this term, "direct fulfilment," for the content of immediate realisation.

CHAPTER IX

LOGICAL INTERPRETATION: MEDIATION THEORIES

§ *1*. *Actuality Theories : Intellectualism*

1. The process of mediation upon which knowledge proceeds has been described by many writers.[1] It issues in the establishment of the objects of knowledge in certain spheres of existence and under a restricted control. In each case, there is a coefficient or sign attaching to the mediating presentation or idea, upon which its assignment to one or other of these spheres proceeds. The idea of an orange mediates the real external thing. Thus systems of realities are built up, toward which the knower is led to take attitudes of acceptance, acknowledgment, and belief.

So far as this system of mediations is concerned, the same is true of values or worths considered as existing and actual. The idea of a present fact mediates the satisfaction associated with it and reached when the thing itself is attained. The idea of the orange mediates not only the actual orange as a thing, but also the pleasure of eating the orange. The coefficient of existence is the same for the thing as for its predicated worth.[2]

[1] See *Thought and Things*, vol. ii., chap. xiv., and vol. iii., chaps. iii., iv.

[2] *Ibid.*, vol. iii., chap. v., § 2.

By the actual, then, we mean a mode of reality reached by processes of actualising; and these processes are those of mediation through ideas. Its most elaborate and explicit form is reached in the discursive processes of the logical mode, which proceed by experimentation and proof. The astronomical body is made actual when its possible existence, anticipated by the hypothesis which states the mediating ideas or facts, is experimentally confirmed or logically proved.

2. It is true, however, that all knowledge, all cognitive apprehension, not merely that of logical process proper, involves the process of mediation more or less explicitly. A thing recognised as what it is and nothing else is always made up or constructed, "actualised" in a word, in a control which makes of it a fact or truth. As actual, it is always a "what," never a mere "that."[1]

The sphere of existence to which it is referred is accepted as proper to it; and the stage, higher or lower, at which it is found depends upon the relative development of the mind. It ranges from the presumption of mere reality-feeling to the assurance of explicit judgment and conscious belief.

On this showing, theories which interpret the world in terms of the actual, limit the real to that which is accepted with some show of confirmation; the real is

[1] See *Thought and Things*, vol. i., chap. iii. It may be objected that our most intimate sense of actuality is present in cases of direct experience or immediate presence, when there is little or no cognitive process. This is in a sense true; but when it is true, the case is one of immediacy (to be treated below), in which there is for consciousness no question of actuality or lack of it. In immediate experience we do not actualise or make actual; we merely find what is present, by contact with it. In this connection, therefore, we restrict the term actual to that which is actualised, that which is taken up, accepted, or found to be actual.

that which is established in a control—external, inner, etc. They limit reality, in other terms, to the outcome of knowledge. These we may call "actuality" theories.

3. One may, of course, go further and prefer one mode of control and one type of knowledge to others; and so restrict reality to that which has a certain sort of actuality. The theory then becomes more special, as in materialism, or again in spiritualism, or yet again in rationalism, and in intellectualism understood in a narrow sense. But for our present purposes, these may all be classed together as theories based upon the same presupposition: the presupposition, namely, that the attitude of acceptance which acknowledges reality is peculiar to those functions by which an idea or other mental content is confirmed in a control or sphere of existence. This attitude begins with the mere presumption of the child's trust and the savage's credulity; becomes belief when judgment arises to reaffirm and extend the objective system of things; and later on lapses into the presupposition that lies behind those operations which extend the established system of knowledge. It is an attitude of knowledge—of confidence, confirmation, assertion — as opposed to that of question, assumption, hypothesis. The actual — whether physical thing, economic value, moral situation, religious truth—is something grounded in observation, demonstration, or knowledge. The statement "it is actually so," is the end of dispute.

Applying the term intellectualism to this group of theories—seeing their exclusive appeal to the processes of knowledge—we may now enquire whether they give a satisfactory account of reality; whether, that is, the interpretation they present is successful in adjusting all the genetic motives of apprehension.

§ 2. *Examination of Intellectualism*

4. Before considering the special forms that intellectualism takes on, we may point out certain general difficulties under which it labours when considered from our genetic point of view.[1]

(1) It is clear that to accept as final the view that all reality is present in the actual, as reported by the intelligence, is to deny to all modes of function other than that of knowledge the rôle of reaching or enjoying reality. This denial extends, of course, to the function of idealisation, by which further value is attributed to the existent or actual and the true, over and above their properties of actuality; as, for example, the ideal values attaching to moral and religious objects. It also refuses to recognise as real certain contents of immediate experience, in which we seem to attain a direct realisation or assurance. All the revelations, in fact, of feeling as such have to be submitted to the discursive tests of knowledge. Recognising that in these cases we are dealing with values, this criticism amounts to the statement that the theory of intellectualism does not provide for the reality of a large group of values.

5. Such an intellectualistic attitude could be justified in this regard only if some motive appeared in the development of the object, in the course of experience, by which the functions of idealisation and direct realisation were subordinated and made instrumental to knowledge or thought.[2] This is the claim made by a

[1] In what follows the attempt is made merely to show in what respects the theories considered meet, or fail to meet, the conditions imposed by our genetic method and suggested by our results.

[2] As would be the case if all appreciations as such could be rendered in judgments of truth or fact. *Cf.* the discussion of Urban, *Valuation. etc.*, chap. xiv.

certain form of rationalistic theory. But we find, on close examination, that the processes of mediation of values as ends, through ideas taken as means, and indeed the entire organisation and logic of interest, are dynamic and autotelic processes.[1] They are not submitted to the domination of the intelligence. They do not rest content with the acceptance of fact, or the discovery of truth by processes of reasoning. So far from finding their demands fulfilled in the establishment of the actual, they go on to erect ideals and establish norms which transcend the categories of thought. It is only as the scaffolding of relational statement falls away, as in the realisation of God in the religious life, that the independent worth of the ideal for feeling and will is discovered.

6. (2) Not only do the active processes refuse to submit to the yoke of the intellect, but analysis reveals positive elements in all our worlds of reality—physical, moral, social—which are not cognitive: elements seeming to come immediately from direct contact with things, or seeming to be achieved only by effort and struggle. We have elsewhere[2] shown that the limits of knowledge are reached and passed in experiences of singularity, on the one hand, and in those of universality, on the other hand. The singular escapes generalisation and resists relational construction; it comes as an immediate realisation. Every experience of exclusive interest—the child's kiss, the drunkard's cup, the image present to the gaze of the devotee—gives a sense of reality more intimate than that of all the proofs of logic.

[1] See *Thought and Things*, vol. iii., chaps. vi., vii.

[2] See the summary given in the chapter on "The Embarrassments of Thought," in *Thought and Things* (chap. xv. of vol. ii.).

Further, the universal and necessary can not be proved to be real; they are assumed or postulated to be so, by a movement which establishes an ideal, even in the realm of the actual.[1] Moreover, the universals of the different spheres of truth, conduct, religion, etc., become relative to one another as soon as these spheres are considered as constituting a whole of reality.[2]

7. (3) A further objection to the theory which considers knowledge as the exclusive organ of the apprehension of reality, appears when we examine the type of construction produced by knowledge or thought. It renders those relations of identity, similarity, and recurrence which take on the form of generalisation. Even the identification of the singular object as the same requires a generalisation of its recurrent appearances in the mind of the one person.[3] This requires and assumes a certain fixity and constancy in the data of knowledge; for the confirmation of the results must be possible, both by the one person and by all. In the result, we have the notion or concept, which is in its intent, and in the verbal symbol which expresses it, fixed and invariable, however much the cases it covers may vary among themselves in other respects. Only on this assumption of fixity and constancy can the processes of discursive thought go on. The truth reached by such knowledge is therefore of a conceptual and relational character, reporting only those aspects

[1] It is only by reason of the historical confusion of "reason" with "reasoning" or thought that the immediate intuitions of pure and practical reason are claimed by thought at all. If we should use the term thinking always instead of thought, the confusion which arises from our denoting universal and necessary principles of reason by the term "thought" would become evident.

[2] See *Thought and Things*, vol. ii., chap. xiii., § 7.

[3] *Ibid.*, vol. ii., chap. x., § 6.

of the real which can be summarised and communicated in verbal symbols.

We have seen that this relational and abstract character of knowledge does not vitiate it; at the worst, it merely renders it partial and in a certain sense remote from the data of immediate apprehension.[1] Even the advocates of radically pragmatic theories admit that knowledge does serve its purpose, as an instrument of life and practice. It is then so far valid for the purposes of the active life, even in cases in which its limitations are recognised. Whatever reality may be in its completeness, it must have aspects, as it would appear, which are capable of taking on the general and conceptual forms of knowledge. This is brought out more fully below.

8. The general point is well taken, however, to the effect that knowledge does not exhaust our apprehension of reality. The singular and immediate, the contents of affective and active consciousness, do not lend themselves to its processes. Reality in certain of its modes does not present itself in isolated "cases," or recurrent bits, among which relations may be discovered or established. On the contrary, it presents itself as flowing, moving continuously, having a dynamic motive in its progress. In the inner world, the self is not a series of detached experiences, but a persisting presence always moving and changing, although always apprehended as the same. Its marks are those of a continuity which shows qualitative change; the opposite of the discontinuity and qualitative identity of the objects established by thought. To the development of practical meaning in general, therefore, it is found impossible to apply the conceptual machinery of cogni-

[1] *Thought and Things*, vol. iii., chap. iv., § 2.

tive logic. The classes reached are too indefinite, the conclusions too indecisive, the mutations too facile, and the contents too relative.[1]

These properties of the affective life are so well known that intellectualism itself cites them in favour of its thesis. Feeling and action, it is said, are too unorganised and irresponsible to report the real. But such a conclusion follows only on the assumption that reality is of the fixed and static nature suited to intellectual apprehension merely. In case of the contrary, or in case of a partial unfitness of reality for intellectual categorising, it may be that feeling, under certain conditions, gives us part, and the best part, of the real. In that quality in which reality does not stand still long enough to be photographed, or does not recur plainly or frequently enough to take on the moulds of thought, feeling alone could secure our contact with it.

9. (4) Another limitation upon knowledge as such appears in the fact that all knowledge is in its nature "common," in some sense social, as has been abundantly shown in detail.[2] It is common in the sense of being open to more than one knower. This is of interest at once for the theory of knowledge and for that of reality. The common character begins with the aggregateness of different persons' perceptions, and advances by stages up to the completely "synnomic" community of judgment and theoretical reason. Knowledge differs from feeling and will in this, that its common strain of meaning is reflected in the object itself whose very constitution and control give all men the right to find it to be what the one person declares. It is part of its intent to hold for the experience of all; and it is not

[1] See *Thought and Things*, vol. iii., chap. vii.

[2] See *ibid.*, vol. i., chap. vii., and vol. ii., chap. iii.

knowledge unless it has this intent. The control of knowledge establishes the trans-subjective reference; and externality once established, repeated perception of the same thing or object is guaranteed.

If this is true, cognitive processes can not report that which is essentially private, personal, intimate.[1] Any aspect of reality which is not capable of being publicly inspected and found to exist independently of subjective interest and choice, is beyond their reach. In this again an advantage is claimed for knowledge. This quality is openly exploited by its advocates; its results are free, we are told, from personal caprice, private interest, subjective illusion. By it reality is once for all defined in disinterested and impersonal form.

10. This is very true; but it again assumes that all reality is of such a type. It denies or neglects those marks of reality which appear in the privacy of the single person's appreciation. It treats consciousness as a sort of reflector of a reality which is apart from it. The reflector is good because it gives consistent pictures of the objects exposed before it. The question of the reality of the mind itself is resolved into this—how far it too can be accurately reflected in the mirror of knowledge.

We find reason to think, however, that the inner world is real in that it fills an intimate and vital rôle of its own. Its primary apprehension of itself seems to be direct, not reflected. In the inner realm there is a series of immediate contacts, revealing values, serving

[1] *Thought and Things*, vol. ii., chap. xiv., §§ 4 ff. We have seen that the generalisation of recurrent affective states in dispositions and attitudes lacks the external points of reference by which judgments of fact are socially confirmed. (*Ibid.*, vol. iii., chap. vii., § 2.)

appreciations, establishing goods, which reveal something more than the properties of the actual given by knowledge. To be sure, knowledge does lend itself to the task of making general and public, in judgments of right, taste, etc., the values of the private life, but vainly, for the most part. And back of these inadequate renderings, the entire inner world, with its first-hand values, its felt persistences, its active efforts, its moral goods, remains a standing refutation of the monopolistic claim made on behalf of knowledge in the intellectualistic theory.

In the logical stage of knowledge or thought, in which conscious reflection appears, it would seem that the entire self might be finally apprehended by knowledge. We do commonly say that we know the self, comparing it possibly with our other past and dead selves, or bemoaning its shortcomings in the moral order. But apart from the limitations actually present upon this sort of apprehension,[1] we find that here still the entire function has the presupposition of the inner sphere in which it takes place—the background or theatre of knowledge itself. This remains true even for the pure presentational or Herbartian theorist. Whatever influence the transition from direct to reflective knowledge may have upon our interpretation, it would be in the direction of introducing a factor of subjectivism into all reality; it would not tend to place new emphasis upon knowledge or thought as such. The inner life is given to us as a mode of actuality; but it has the nature of an immediacy, and as such it underlies the mediating processes of reflection.

The objective self that we judge, describe, and esti-

[1] See on the "Apprehension of Immediacy," *Thought and Things*, vol. iii., chap. xiv., § 6.

mate is mediated by various partial and detached ideas; as object of reflection, it is a construction. Over against this, however, stands always the subject of reflection, an immediate process, a continuing inner control.

11. All these criticisms of intellectualism, and of the view of actuality that goes with it, seem to come back, finally, to the analysis of knowledge itself. In every case of knowledge, there is along with the external and convertible skeleton of fact or truth, an element of interest, a personal intent, which is a constituent part of the whole. The mere fact, the bare neutral truth, is always an abstraction, serving to set the element of external control apart from the appreciating and determining interest, without which the object would not be what it is. But even for purposes of abstraction, the other pole of the knowledge relation is in evidence. Over against the external fact, the inner life reveals itself. In order to establish the neutrality of the object, science ignores the mass of pertinent worths. In this way science makes its own arbitrary measure of reality, and estimates all reality by this measure. Of course it finds what it looks for.

12. It should be noted especially, however, that these arguments, admitting their full force, do not disprove the validity or impair the value of knowledge. At most, they limit its scope and define its rôle. The advocate of actuality still has his feet firmly planted on the solid ground of the control found in knowledge. He says rightly that all confirmed reality, of whatever kind, all controlled meaning, of which common acceptance is justified and upon which common action is possible, is reached by processes of knowledge. Genetic analysis shows that the credulities of faith, the illusions

of imagination, the artificialities of formal argument, the vanities of misplaced hope, the excesses of passion, can be detected and cured only by the sober experimental methods proper to judgment. Where would society and the individual alike be if the dream, the vision, the postulate of the "will-to-believe," even the ideal, were not controlled in theory and practice by the reasonable insight and common proof of thought? We do not find here justification for the extreme doctrine of the a-logism which charges knowledge with rendering only a conceptual distortion of the real, and finds in mediation through ideas only a means of banishing the actual and concrete. On the contrary, the end of mediation is to restore the actual, to bring the imaginary within the range of concrete control. Without the sifting and testing processes of knowledge, it is difficult to see what resort there would be against the delusions of the insane, the utopias of the visionary, and the nightmares of the dreamer.

Criticise knowledge, therefore, as we may, as being static, rigid, artificial, geometrical, we can not dispense with it. It makes an essential report on the nature of things. Without it, we might have immediate contacts with reality, as many have claimed; but the reality given in such contacts would be, as we will have occasion to see later on, without meaning to instruct or ideal to guide us.

13. It would seem, therefore, that while knowledge gives us an aspect of the real, it does not exhaust reality; that the entire series of aspects given us as immediate values, ideal constructions, and direct and intuitive apprehensions, unless considered as in some sense secondary and illusional, must have some other justification in our theory.

14. Of the actuality theories to which these general criticisms apply, the narrowest is the Positivism which accepts only the control secured by positive experiment, excluding the processes of rational and discursive proof, so far as they can not be tested by fact. But this is to set unnecessary limits to knowledge, since all properly conducted argumentation is in its nature experimental;[1] and the demand that this mark shall be explicitly formulated in the method is to deprive the scientific man himself of his most affective weapons of proof.[2]

Materialism goes a step further. It is refuted, without more ado, by the fact that spiritualism rests upon the same sort of assumption that it does: the assumption, namely, of the reality of one sort of actuality together with the denial of the other and correlative sort.

Spiritualism, in certain of its monistic forms, is little better off. For unless considerations of other sorts than the grounds of actuality are taken into account, it can not disprove the physical or annul it, any more than materialism can disprove or annul the spiritual. Both represent forms of actual control, which are always present together and which divide the territory of fact between them. In fact, however, the grounds upon which spiritualism is usually defended are broader; one attributes to the soul more than simple actuality,

[1] Shown in detail in vol. ii. of *Thought and Things*. *Cf.* the recent article of Rignano in *Sciencia*, 1913, N. xxvii–1.

[2] It is difficult to define positivism, since it has become a method and a temper of mind rather than a theory. The present writer's work, *Thought and Things*, has been described as neo-positivistic (see Roberty, *Revue Philosophique*, Jan., 1914, pp. 21 ff.); and in method and certain of its results it is so. But in the results, which deny adequacy and finality to knowledge as such, and find reality to be progressively realised in æsthetic contemplation, positivism of the nineteenth-century type, at any rate, would hardly recognise itself.

finding in it moral and ideal properties. The soul is considered as a value, and to be real for that reason. This places the theory of spiritualism, in certain of its forms, among those of the "idealising" class which are now to be considered.

§3. *Ideality Theories: Voluntarism*

15. The reservations now made as to the rôle of knowledge in revealing reality are, it is plain, largely in the interest of will and feeling. The will postulates values desired, ends pursued, ideals cherished; and feeling brings us into contact, directly and without mediation, with aspects of the world which we can not express in the relational terms of knowledge.

The development of the body of "teleological" meaning, constituting worths and ends, reveals motives at work correlative to those of knowledge proper.[1] The process by which the actual is discovered or confirmed is matched by one by which reality is assumed, supposed, or postulated. The imagination idealises, while the judgment actualises. The imagination goes before the judgment, schematising, prospecting, experimenting, treating its constructions as being real, as-if-real. The ideal is the world of the "as-if."

It is the significance of this as-if-real—of that which is assumed, supposed, postulated—that concerns us here. Is it no addition to the actually real?—is it without significance for the "make-up" of things? Is it only a fanciful appendix to the actual, or only an imaginative reproduction of it? Does the recognition of the worthful side of existence, of the possibilities

[1] Cf. *Thought and Things*, vol. iii., chap. v ff.

of an ideal sort resident in the whole, take us no further in understanding its intrinsic nature?

16. In reply to these questions, the theorists of a second group, the "idealisers," speak with emphasis. The interest of idealisation, they say, is as genuine and as urgent as that of actualisation. The worth of a thing is as real as the fact of it. Indeed, actuality itself is but the ideal of one of our interests, that of establishing neutral fact or truth. Actuality, like other aspects of objects, is fundamentally a value; like the others, it is an end or ideal. We reach a neutral objective world of fact or truth because we have an interest in doing so: an interest in disengaging the stable, external, and common from the changing, internal, and private in our experience. At the best, the whole system of truths is instrumental to the values of life, found in morals, religion, art.

Looked at broadly, therefore, it is the body of interests, and the self that exercises these interests, bearing in its bosom the system of ends and ideals which fulfil them, that impress these minds—the believers in the importance of idealisation and of value. The emphasis goes over to the inner life, the self, the control within. The external fact or truth becomes merely one of a series of postulated values; it is real only as it ministers to the development and realisation of the self. The truthful is to be read as a partial fulfilment of the whole ideal, which alone is the whole real. The ideal—the term in which reality finally and fully embodies itself—involves value.

17. Among the views falling in this class, we have all the systems of idealism—defined as systems based on "ideals"—such as Platonism, certain spiritualisms, moralism, various sorts of voluntarism and, on more

modest grounds, pragmatism and instrumentalism. To all these theories alike, knowledge is not an end in itself; it is merely a means. Truth does not realise or summarise reality; it reflects it and makes it available in a certain form of value. Actual existence is not the final form of reality, but at most a stage in its movement, a phase only of its series of appearances. Only the supreme value, the ideal good, is absolutely real; all other truths and values alike—particular facts, goods, beauties—are but shadows of it, approaches to it.

Seeing that it is for the will that all value takes its rise, and also that the ideal is that which is set up as end to the will, the historical term "voluntarism" may well be applied to this general class of theories. As actualism, setting out with the type of reality achieved or presupposed by knowledge, reaches Intellectualism, so idealism as thus defined, setting out with or postulating the reality achieved by will, reaches Voluntarism.[1]

§ 4. *Examination of Voluntarism*

18. In proceeding to criticise such theories, we should bring out clearly their distinguishing mark. It is not value considered merely as fact, as attributed worth, value that is known and rendered in judgments of appreciation or truth, that they have in mind. For such values, considered as part of the actual, are the property of knowledge. The good taste of the apple

[1] We are concerned with the general theory of idealisation, however, under this designation, rather than with the more restricted theories of voluntarism strictly conceived. There are certain advocates of ideal reality, spiritualists and theists, who would not consent to be called voluntarists in the metaphysical sense.

is a value; but it is also a fact. It is realised in sensation and generalised in recurrent experience.

It is, on the contrary, the value that is unrealised, assumed, ideal, and all values so far as made ends of will, or idealised; the worths which stimulate the will and exercise the imagination as schemes of greater good, the goods that tempt and the purposes that inspire—it is this realm of values which these theories would make real. The intellect decrees, "let the true be real in its neutral existence, apart from interest and will"; but the will replies, "let the good be proved true or not, let the ideal be realised or not, still the world is one of goods, pictured in imagination, embraced in volition, postulated in the ideal. Its significance is for the self, and its reality consists in its power to satisfy." In the function of volition the real is approached; its character disclosed. The fundamental human interest is not theoretical, but practical; the fundamental real is not the true, but the satisfying, the perfect.

Certain of the more sharp and decisive criticisms of this theory may be stated.

19. (1) Granted, it may be said, that knowledge is one interest among many, and that its results represent one ideal; still if an ideal is to be taken as the fundamental reality, why not this one?—the ideal of truth? To deny its reality is to open the way to the denial of all ideals, for they all rest upon the same basis of postulation. If the postulate as such is to be taken as the embodiment of reality—since it satisfies our will and fulfils our interest—then surely the postulate embodied in the ideal of truth can not be left out or considered secondary to any other. And this for certain reasons.

(2) The cognitive mode of process seems to have arisen genetically to bring coherence, definiteness, and control

among the various postulates of the active and emotional life. The rôle of the discursive and experimental processes is to confirm and establish what is otherwise hypothetical, the matter of assumption and postulation. The task of knowledge is to turn assumption into belief.

Without knowledge, indeed, desire is undirected, will is capricious, action is without confidence, imagination is fanciful, morality is hesitating, intercourse is ambiguous and unfruitful. With it, on the contrary, under its ideal of truth, desires are co-ordinated, will passes into rational choice, action becomes confident, imagination merges into judgment, and morality frankly meets the other person and God. In short, all the ideals become intelligible, continuous, and co-ordinated goods only by reason of the function of knowledge, which not only discovers the ideals, but enforces them *by finding them true.*

(3) The ideal as such, the good in all its forms, and with it the self whose interests it satisfies, develops only through the mediation of knowledge. All idealisation takes form through the actual. This is the element of truth in the saying of the voluntarists to the effect that knowledge is instrumental to value, theory to practice. While this is true, still the relation may be read with equal right in the reverse sense; the sense, namely, that values or goods depend upon truths, upon knowledges. This is illustrated in the "Socratic paradox," "no one can knowingly do wrong"; for it is equally true, on the same presuppositions, that no one can ignorantly do right. The will must be informed, in short, the character shaped, the imagination directed by the true, if ideals are to have significance for the self that espouses them. Knowledge comes through

external control, but self-control comes through knowledge.

(4) The body of stable facts and truths, employed as means to the achievement of ends, is necessary, if the self is to grow from the stage of instinct and impulsive action to that of the reflective postulation of ideal goods. Not only are the ideals themselves reflective elaborations of truths; but the very self, whose interests are cited as motive to the postulation of the ideal, is the product of the conditions of actuality, physical and social, which produce and in turn satisfy its desires. This does not in itself reduce the ideal type of reality to the actual; but it does show that the actual, in the shape of the self, is necessary to the very postulation of the ideal.[1]

20. To sum up these criticisms, we may say, in a word, that idealisation, considered as resting upon will and issuing in the postulate, can not even prepare its own proper object without assuming the validity of knowledge. Knowledge is essential; actuality and truth give body to the reality postulated in the ideal. This is true of any ideal whatever, so far as it incorporates genuine worth—the ideal self, the moral good, God. The dualism of means and ends remains to the last to prevent the acceptance of any form of voluntarism which refuses to give to the actual its share in the territory of the real. The means must be actual in order that the ends may be truly ideal.

We must conclude, then, that, while intellectualism can not account adequately for reality, voluntarism, understood in the sense explained above, is equally

[1] In view of this, many voluntarists go over to some form of met-empirical will which identifies itself with the self: *cf.* Urban, *Valuation*, etc., chap. xiv.

unacceptable. They represent principles each imperatively demanding recognition, but each obstinate to any final subordination to the other.

21. The theories based on idealisation take on many forms, most of them departures from a pure monism of the ideal. Voluntarism in the narrower sense often combines, with the postulate of the ideal end, the acceptance of a realistic and immediate doctrine of the self. Such a self is revealed, it is said, in experiences of effort, free choice, and interest. This latter doctrine, considered as a form of immediate spiritualism, is taken up again later on. It does not properly give support to voluntarism, as against intellectualism, since intellectual events, considered as experiences, are as immediate in consciousness as volitional; and since the self of one interest is as directly present as the self of another.

The historical forms of voluntarism, however, from Saint Augustine to Maine de Biran, have found, in the consciousness of voluntary effort and activity, evidence of the reality of the spiritual principle—evidence not afforded, that is, by intellectual activity. On our interpretation of both functions as motived by interest, however, this claim is not well founded.

22. In recent pragmatic views, an instrumental theory of knowledge is combined with an alogistic theory of reality.[1] But they do not appeal to the ideals of truth and value as giving final form to reality. In this sense they are positivistic and relative.

In moralism and theism the ideal postulates are defended. An ideal moral and spiritual life is the end

[1] Cf. Heath Bowden. *The Principles of Pragmatism* (1911), in which a radical voluntarism extends itself in a "dynamic" theory of nature.

of all existence. Logically this leads to a form of personal pluralism in which reality is taken to be of the nature of a community of wills or personalities; or to a theism in which the ideal is realised in the person of God.

23. All the objections stated above to a view which makes knowledge merely instrumental and truth a means rather than an end hold against such a moralism. If the self of the moral life is an actual self, reached by processes of knowledge, then the cognitive function is indispensable; and its ideal, that of truth, must be recognised. But this carries with it the reality of the external world. If, however, the opposite is the case, the self being reached by immediate feeling, then reality remains at the stage of assumption, postulation, and intuition.[1] In any case, to such a view the whole problem of the existence of the external world remains over; as well as that of the sort of community or intercourse possible among the real selves postulated as the more or less isolated centres of moral interest.

24. As to the theistic postulate, we have already seen its limitations as arbiter of the real. It attempts to combine the ideal and the actual in a real person infinite in his nature. But so far as ideal, God is a postulate, an assumption, beyond the grasp of intellectual apprehension. So far, on the other hand, as God is actual, he is not ideal, but remains a finite person, a social fellow, along with other persons. He is the great Companion. As such he could claim no reality different from that of others, nor from that of other things whose existence is also established

[1] One of the views based on immediacy which are to be discussed further below.

by the processes of knowledge. The social demand, active in the religious life, is for a personal helper, a *socius*,[1] and in God this sort of reality is said to be realised. Yet the ideal of personality set up by the spiritual and religious postulates is unapproachable and eternal, not social in any concrete sense, since not proved to be actual. The religious mode of realisation can not of itself escape this antinomy.

[1] I find in an account of the first Gifford lecture of A. J. Balfour (*Public Opinion*, Jan. 16, 1914), the lecturer reported as saying that the God of which he would speak was "a spirit communicating with other spirits . . . what he should not think it profane to call a social God . . . to be distinguished from the sort of Absolute to which everything . . . was indifferent . . . etc."

CHAPTER X

IMMEDIATE THEORIES: (I) THOSE BASED ON THE PRIMITIVE AND THE TRANSCENDENT

§ *1.* *Their basis, Immediacy*

1. In the movement of conscious experience there are, as we have seen, certain well-marked periods. The earliest period is that in which the distinctions have not yet clearly appeared which develop into the dualisms of thought. The primitive immediacy of feeling is largely sensational, and passes, by means of the twofold reference of conscious states, into the mode in which the external and internal worlds are together mediated. Knowledge gradually frees itself and becomes the instrument of the mind.

A second great period is thus brought in, that of mediation through the two contrasted controls. In this period the processes of active and voluntary mediation are developed. The contents of knowledge become means, in the pursuit of ends. Values are set up, ideals erected, ends pursued.

This second period is given over to explicit mediation, both cognitive and active. It reaches its extreme development in the logical or reflective mode, in which discursive processes become the tool by which remote conclusions are brought within the range of reasonable thought and judicious practice.

In the next, the third period, a second immediacy appears. The ideals and ends erected in the pursuit of mediation take on unconditional form. The principles of reason and the norms of practice, as we have shown elsewhere in detail,[1] come to stand alone each for itself as the outcome of its own process of mediation. The categories of thought and the rules of pure reason attain a formal universality which dispenses with empirical content. They cease to be means or instruments, and become independent ends and unconditional truths. Likewise, the norms of practical reason attain formal universality and necessity and present themselves as self-imposed values, no longer requiring empirical or prudential means.

In both cases, the theoretical and the practical, the process of mediation abolishes itself by attaining its end-state in a new immediacy. In one case, it is an immediacy of rational, in the other of practical, intuition.

2. At this stage the immediacy reached is secured as the outcome of the logical and practical processes of mediation operative in the earlier stage; it represents these processes at their completion and fulfilment. But it remains in its character twofold. The pure reason and the practical reason—or the two states of mind designated respectively by these terms—are not the same. The spheres of life to which their respective categories apply are different. Both in origin and in application, one is intellectual, rational, the law of thought; the other, both in origin and in application, is practical, moral, the rule of practice. This has been the result of our investigation of the origin of the pure and practical reason.

[1] *Thought and Things*, vol. iii., "Interest and Art," chap. viii.

There are thus two periods of immediacy, the one primitive and a-logical, the other transcendent and hyper-logical. With the one, conscious process begins its career; with the other, it ends it. And between the two lies the vast tract of mediate function, which ministers to the life of neutral knowledge and strenuous practice.

3. In our detailed study, however, and especially in one of our more special expositions,[1] we have found, besides the two sorts of immediacy just described, a third, that of a mode of function in which consciousness, in the very midst of its dualisms and mediations, seems to wish to forget or to banish the distinctions and oppositions in its content, and to bring its forces into equilibrium in a mode of direct apprehension—to throw away its lenses and look directly into the face of reality. This is characteristic of the semblant, and more especially of the æsthetic, imagination.[2]

In this it is possible to find a synthetic or reconciling immediacy which presents certain contrasts with the other two. It does not belong, as they both do, to one place or stage only of mental development; but extends throughout the whole movement. It is a mode of realising, as opposed both to cognising and to idealising.

4. It is plain that each of these three immediacies may serve in turn as basis of interpretation. All such interpretations would, with equal right, be termed "immediacy" theories. We find that each of the three has served in this rôle, in the history of philosophy. There are theories based on primitive immediacy, on transcendent immediacy, and on immediacy of syn-

[1] *Thought and Things*, vol. iii., chap. xiv., §§ 5, 6.

[2] See *ibid.*, vol. iii., chap. x.

thesis or reconciliation. The first two may be described one as primitive "affectivism" the other as "transcendentalism"; and there are many varieties under each. Under the first, we find Mysticism, Sensationalism, Immediate Realism; and under the second, Intuitionism, Absolute Rationalism, Higher Mysticism.

The third type is not so clearly defined in the history of thought, and has no recognised name.[1] The name varies naturally with the mode of function to which the synthetic or reconciling rôle is assigned. To us, as will appear in detail below, it is the æsthetic function—hence the term Pancalism, already suggested for the special form of the theory stated below.[2]

§ 2. *Theories based on Primitive Immediacy*

5. The assumption that in a certain direct awareness consciousness has its closest contact with reality, has always been attractive, both to those who are not speculatively disposed and to those who have exhausted the resources of speculation. The primitive data of consciousness are, of course, in some sense the foundation of all further data, and the return to immediate certainty is a source of consolation after speculative misadventure.

Reserving for the present the theories that resort to a higher or hyper-logical immediacy, of the nature of

[1] If one should seek for a term in "ism" analogous to those in use in other cases, no doubt "contemplationism" would do, for the word "contemplation" is already widely in use for the act of realising a content directly, an act having both cognitive and active factors. In view of the close resemblance, however, of all the sorts of immediacy among themselves, considered as states of feeling, I shall employ the phrase "constructive affectivism."

[2] See the preliminary sketch given in *Thought and Things*, vol. iii., chap. xv.

intuition, let us enquire how the awareness of primitive or undeveloped experience may be turned to the advantage of theory.

6. (1) Primitive Mysticism. In simple and naïve mysticism, the appeal is made to an "inner light," a source of direct illumination. It is accompanied by states of ecstasy, induced by withdrawal from the world, giving freedom from thought and care. In certain psychosophic systems, a detailed or long-continued technique is resorted to, to deaden the functions of thought and dampen the stirrings of will. Thus made vacant, the soul may be filled with an inpouring of reality. Oriental mysticism is of this type; it forswears speculation, and counsels isolation and personal absorption. The divine is reached by a contemplation which is the more profitable as it is the more vacant of intelligent interest and free from the fretting of desire.

Its justification as a point of view has remained, like itself, unspeculative. It is a spontaneous way of treating life and the world, a first interpretation, externally allied with the social and religious interests which take root in it. The fervour of primitive religious ceremonial and the effervescence of social excitement terminate in such states of mind. They thus take on worth as being symbolic of the values of communal life and of the mysterious things of the occult world. Personal thought, intelligent will, ritual observance, morality itself, are all swept away in the onset of the mystic efflatus. The result is a purging of the mind, a physical fatigue, a consecration of spirit, a sense of confirmed faith and new inspiration. The real is *realised* in the mystic trance. The prosaic world is a mere dream, its affairs a hindrance, its interests a distraction; the real lies beyond it all.

7. The reason that this point of view has remained without theoretical justification is probably twofold: it both has none and needs none. It needs none because, to those who accept it as final, its evidence is just in itself; it is by nature apart from demonstrative processes of proof. It has no proof, in the strict sense, because its data are primitive; there is nothing more fundamental or simple to which they can be referred. It is, in short, just the merit of the primitive that it is realised only when discursive and voluntary interests are in abeyance or are undeveloped. To get back to the primitive is to strip one's mind of the mass of logical and sentimental over-growth.

8. It is for this reason, however, that the mystic point of view, while so recurrent, has remained so unproductive. The return to immediate life, to the intimate flow of things, as is enjoined by James—to cite the latest view in which the primitive has had its claims upheld—affords rather an effective protest against intellectualism than a position having logical ground. To come close to unordered sensations, to realise a "blooming confusion," to feel the *élan* of pure duration, is not to apprehend a very extensive reality, whatever we may say of its spicy quality. And the very qualitative distinctions themselves upon which knowledge proceeds and practice depends, taken in their primitive simplicity, are mere contacts with this or that, without further meaning. If this were all, human commerce with reality would be about the same as that reached by low organisms, except perhaps for the addition of certain new qualities; and however pungent and moving the new contacts might be—the moral, the æsthetic, the religious—it would still seem that the intellectual and practical machinery of the

mind must have arisen to fulfil some useful function. It is here in fact that the proper theory of the immediate has to begin its construction, either by reading something into the primitive or by extracting something from it. In either case, it is not its primitiveness and simplicity that makes it available for theory, but the contrary.

9. As soon as mystic theory began to take on positive character—as it did in Plato, Boehme, and Plotinus—we come upon one of two alternatives. Either it goes over from a primitive immediacy to one that is very developed, transcendent in its type; or it finds in the immediate certain complications or implications which destroy its simplicity. It is especially easy, from our knowledge of what develops later on in growing experience, to reconstitute immediacy as a mosaic of hidden motives and potencies.

10. (2) Subjectivism and Idealism. The first advance in this direction was that which isolated the subjective point of view as such. This became the safe possession of the Greek thinkers in and after the Socratic period. Becoming more and more clear in Plotinus and St. Augustine, it became in Descartes the presupposition of reflective thought. Modern philosophy has never seriously disputed the principle that reality makes the revelation of itself in conscious states—sensations, perceptions, ideas—and materialism itself justifies its negation of mind by citing *what we know* of matter. We know matter in motion and we infer that its properties are such that our minds and our knowledge are secondary and derived products. Even should we admit such a contention, the fact would remain that such a thing as knowledge is present in the real, and that it is only by appealing to knowledge that the

unreality of mind can be proved. The presupposition of conscious process remains intact in any case.

On this presupposition repose, in fact, the systems of philosophy classed together as idealisms.[1] Subjective idealism is the extreme form of the theory which develops the subjective postulate. Reality is construed in terms of a monistic subjective principle.

11. The difficulty arising from the emphasis placed upon the presupposition of consciousness, taken in the subjective sense, is plain. While assuming the inner life, one may fail to recognise equally all the processes in which that life takes on definite form. These processes proceed in reference to two opposed controls, inner and outer, the self and the external world. The trans-subjective reference is universal and compelling, all through mental development.[2] The foreign control exercised upon perception by the external, is as real and definite as the inner control exercised through the active life by the self. To assume consciousness, therefore, as it is, in its actual development, is to accept its assertion of a reality external to it.[3]

This point is reinforced by the consideration that in

[1] Understood in the sense of "idea-isms"—systems which recognise the mediating rôle of subjective states or ideas; not in the sense of "ideal-isms"—systems which turn upon the recognition of ideals. This ambiguity already pointed out in the use of the term idealism in English leads to much confusion.

[2] The use of the term "inner sense," for the organ of apprehension of the world within, gives us verbal evidence of this; the very term suggests the correlation with the "outer senses."

[3] *Cf.* the article "Mind and Body," in the *Psychological Review*, May, 1903. As usually put, this criticism drives the subjectivist to "solipsism" —the acceptance of merely the one private mental life.

But even that refuge is properly to be denied to him, I think; for the private person, the one mental life assumed by solipsism, could not mature without involving itself in a mass of inter-subjective and trans-subjective relationships.

the external world, so recognised, other minds find themselves; they are also foreign to the single person's psychical life. And, further, they are apprehended in experiences in which personal selves and impersonal things are inextricably interwoven *inter se.* To deny the trans-subjective reference, while retaining the subjective point of view, therefore, is to cut off the *inter-subjective*—to deny knowledge of other persons or communication with them. Experience remains immediate, it is true, but it becomes also unsharable. It may be primitive, but it is nothing else. It is as unavailable in theory as it would be insignificant in fact. It would be typified by the flash of consciousness that passes—if it does—over the surface of the jelly-fish.

12. Of course the idealists do not stop with the jelly-fish; they find implications which introduce implicitly or potentially a developed mental life. But this procedure, if it follow the leading of nature, issues, not in monistic idealism, but in dualistic realism. For the very form which the hidden implications take on, so far as they are really present, is that of mediation of one type or the other, intellectual or practical. Seeking to remain monistic, we find ourselves on the ground of one of the theories criticised above, the theories of actualisation and idealisation; the idealist, in the sense of subjectivist, becomes logically either intellectualist or voluntarist.[1] In either case, as we have seen, he is bound to recognise, along with the postulate of subjectivity, some sort of trans-subjective existence.

In view of this, most systems of spiritual monism

[1] The third alternative, the resort to immediacy of the higher intuitive or synthetic type, is considered further below.

fail to effect a satisfying synthesis. They take advantage of the immediacy of the subjective to refute materialism and naturalism, and then go on to read into the subjective a positive content in the interest of a mental principle or soul. The difficulty with this is that the properties claimed for immediacy do not warrant such a procedure. The self, the mental considered as positive content, is an organisation effected under a control strictly correlative with that of the external. To get the mental and the social, one must get the trans-subjective also, the physical.

The most interesting way yet suggested of meeting this difficulty is that of construing the external world entirely in terms of the social. This would give what we might call a "communal idealism."[1] If made out, however, it would be a "mediate" theory, not one of immediacy; and it would still have the difficulty of reconciling the modes of actuality and ideality, as already pointed out of other such theories. Is the social order, of which reality would consist, to be looked upon as an actual or as an ideal order? The demand for such a synthesis of actual and ideal, in the sphere of the social, the ethical and religious postulates seem partially to meet, but not adequately.[2]

[1] Such a resort was intimated by Lotze, and has been explicitly suggested by Ormond (A. T. Ormond, *The Foundations of Knowledge*, chap. vii., especially pp. 202 ff.). In principle the conception goes back to Leibnitz's theory of monads whose essence consists in presentation.

[2] The attempts, notably that of Lotze, to sublimate the physical in the social, rest in fact on the ground of the inconceivability of mechanical interaction in nature. But the substitution of social for mechanical interaction is analogical and without real proof. The only way to secure such a position would be to discover a mode of apprehension in which reality, while "extra-psychic" (foreign to the individual) is not "trans-subjective" (foreign to society). We will see later on that in æsthetic contemplation this condition is fulfilled. The work of art,

§3. *Theories based on the Immediacy of Completion or Transcendence*

13. Obviously the immediatist is thus led to pass from the primitive or original to the transcendent or completed, in his interpretation of immediacy. He goes over to the hyper-logical or intuitive, in which the processes of mediation themselves issue. Passing over the movements of knowledge and will, and fixing his gaze on the ideals, the consummations, the end-states of mind, he *realises* reality by innate ideas, higher instinct, pure reason, intuition, the *a priori;* or finds it in a mystical divine presence apprehended by the eye of faith.

Here we have a group of theories for which the immediate data of consciousness are not the "low things," the simple and primitive contacts, here and there, with the flowing stream; but the "high things," the universals of thought, the absolutes of value, which are implicated in experience and make it possible.[1]

Three types of view may be distinguished under this heading: Intuitionism, Transcendentalism, and Higher Mysticism.

14. (1) Intuitionism. The intuitionist prefers the higher immediacy of fulfilment or completion to the lower immediacy of primitiveness. He correctly says

---

having a synnomic meaning, is charged with a communal self, while still remaining for the individual part of external nature. The life attributed to it, in the observer's contemplation, is not one belonging to an individual self, but one through which each individual spectator partakes of a larger life. See below, chap. xiv., § 6.

[1] This has been also the line of actual historical progress. After the immediate subjectivism of Descartes and Malebranche, came the refined spiritualism of Leibnitz and the transcendentalism of Kant; and after the immediate sensationalism of Hume and Berkeley, came the intuitionism of Reid.

that reality, if it is to be of any significance, must be complex and full, not simple and empty. Why realise a something that amounts to nothing? On the contrary, he goes on to say, when we "realise," although we do come in contact with something, this something is found fit to be known, or imagined, or feared, or worshipped; it is a "some-what," having a wealth of connotations, not merely a "some-that," with no connotations at all. So the question arises as to how an immediacy of fulfilment, consummation, or completion, can retain a content which, if present at all, must be due to preceding processes of mediation, and which can continue to mean what it does only because these processes remain, in some obscure way, in operation.

15. The intuitionist, however, does not accept this last statement. Intuition for him is the immediate apprehension of what is of the highest and richest import: the self, the external world, cause, freedom, duty, God. Such high truths come directly from reality into the soul by the avenue of intuition.

As soon, however, as the intuitionist is pressed to point out some item of knowledge surely given by intuition—that is, given independently of prolonged learning and experience—he finds himself embarrassed. Descartes' "innate ideas" are criticised by Locke; the rationalist's native categories of cause and space are refuted by Berkeley and Hume. The familiar issue of the controversy is that intuitionism—except when it defines itself in terms of the primitive, as explained above—finds itself driven into some form of formalism, in which the content of knowledge, the concrete truth which gives value to the intuition, is handed over to experience. What remains of the intuition is a merely formal or logical principle by which the content is organised. Intuition supplies the form, experience the matter.

The criticism to which the intuition theory, considered as an interpretation of reality, is then exposed, is evident. It so divorces intuition from experience that its entire concrete filling or content is lost. Logical principles become tautologies, ethical maxims become formal rules, space and time become vast emptinesses. We have a system of notions which are practically superfluous and worthless. A progressive experience is after all necessary, proceeding by the organisation of knowledge and interest under actual genetic motives. It is only in the outcome of actual experience and through a reflective interpretation of it, that we become aware of the logical categories and practical imperatives by which we conduct our lives. The separation of form and matter is simply an artifice; such and such form is form only as there is such and such content which is so formed.

16. It now becomes clear that if any *raison d'être* is to be assigned to the supposed intuitions, thus left suspended in mid-air, so to speak, it must be by the assumption of some reality existing outside of experience altogether.[1] Accordingly, two further alternatives logically arise, those of dogmatic spiritualism and transcendental absolutism. One of these is historically the alternative of the Scottish realists, the other that of the Kantian and post-Kantian German transcendentalists. These positions are both transcendental

[1] The new intuitionism, represented by Bergson, does not seem to escape this assumption. The time and free will which are considered immediate data of consciousness find their supplement in a creative evolution in which the principle of life takes somewhat the place of the spiritual principle of other theories.

in a broad sense; since for both reality is something beyond experience, although given in an immediacy of the type of fulfilment or completion. Technically, the term transcendental is generally applied, however, only to the German absolutism.

17. (2) Dogmatic Spiritualism. The hypothesis of spirit or soul is a simple and direct resort. Intuition becomes the organ of apprehension of the soul or spiritual principle. The soul reacts intuitively to reality, realising directly both itself and the world.

This, however, amounts to an interpretation of intuition. It gives it a basis, by means of an assumption, instead of merely accepting what intuition can immediately give. The soul or spiritual principle thus supposed is itself a concept or notion, drawn from extended experiences of knowledge and action, of interest and effort.[1] We know the characters of the self, as we know those of the world, through tentative and experimental processes, social in character; through recurrences, conversions, failures, successes, inferences, all shot through with threads of social intercourse. To cite such a notion in support of intuition is to deny the efficiency or the sufficiency of the latter; for it amounts to substituting for it the outcome of the mediate and discursive processes which intuition is supposed to supplant. Once granted the notion of a substantial soul, the presence of intuitive categories or rules may be taken to confirm it; but the reverse procedure is not valid: the intuitions can not be made the basis of knowledge of the spiritual principle, since they

[1] To this it may be replied that the soul is itself an intuition. But even so we may still ask how that would help us to know what the soul is really like, what its actual characters and powers are. These questions can be answered from personal and social experience alone.

are void of concrete content. Apart from experience, they are formal and empty.

18. Moreover once begun such a procedure has neither control nor limit. All sorts of dogmatisms follow. Whatever logical classes, rational identities, practical casuistries, psychological faculties, seem convenient in the progress of apologetic discussion, the soul is big enough to contain them all; they, like it, become intuitive, beyond the reach of proof, as they are also beyond the range of doubt. The development of the Wolffian dogmatism before Kant, and that of the faculty psychology in Scotland, show the extravagances to which such a theory lends itself.

19. (3) Kantian Criticism. The other form taken by this sort of immediatism is more profound. It consists in a critical examination of experience with a view to discovering exactly to what extent there are in it elements of apprehension which are synthetic, not merely analytic, and not due to experience. Do we have realisations over and above the discoveries we make in the exercise of our perceptive and discursive faculties?

Besides answering this question in the affirmative, Kant instituted a comparison of the great modes of mental function in respect to their relation to reality, stating the problem of the comparative morphology, as we have called it, of the formal elements—the *a priori* realisations—of intellect, will, and feeling. Which of these gives the most valid, the "realist," real?—and how do their different realisations go together? Do they harmonise or conflict with one another, supplement or interfere with one another?[1]

[1] This is the problem we are here taking up, except that for Kant it was the comparison of *a priori* forms (transcendent and logical), while

In the outcome, Kant finds *a priori* forms, immediate and intuitive realisations, in each of the three great modes of function. But when he comes to consider their respective validity and estimate their content, he does not stop with dogmatism, nor with mere intuitionism. On the contrary, he establishes the point that rational form is empty, but for the content which comes from experience. The categories of the reason are valueless, apart from their phenomenal use. The appeal to a substantive soul or spiritual principle in support of these forms is not valid, since this principle is real only for thought; it is "noumenal" in its character. To apply the categories to the thing-in-itself would be to bring this latter into the sphere of phenomenal existence and simply duplicate the experience we have. Rational intuition, therefore, is not an independent organ of the apprehension of reality in itself. The categories of thought and the ideas of the reason are merely regulative principles of our apprehension.

The real world and the real self, reached in our experience, are phenomenal; but for Kant they are the only sort of reality that our processes of knowledge are capable of attaining.

20. The practical reason, operative in the sphere of will, contains also its *a priori* form, the imperative of duty. And in it Kant finds the justification of the ideas of reason—God, freedom, immortality—that is wanting to the pure reason itself. Here, in the ideal

our problem is the comparison of all the modes of reality that consciousness accepts or assumes, of whatever kind (empirical and actual). The method also differs. Kant's method is that of criticising a finished outcome: asking how it is possible; while ours is that of describing and interpreting a genetic movement.

of the good, the ideas of the reason become constitutive. The good is an absolute teleological ideal expressive of the nature of the moral agent. From this follows the reality of the soul as a free and immortal personality.

While the pure reason, therefore, can not reach a reality that is more than phenomenal, the practical reason does so, in the postulate of the absolute good. There is thus an opposition between the two, or rather a contrast; since as organ of the apprehension of reality, the preference attaches to will. This leaves the rational nature of reality undetermined, the cognitive processes spending themselves without reaching their goal. Is there, then, it may be asked, any further justification of knowledge as such, and any evidence of essential unity as between the theoretical and practical sides of the reason?

21. This is the important question of the morphology of reality as indicated above. Whether or not rational intuition reaches an absolute, whether or not the practical reason makes good its postulate in any final sense, this problem remains over: that of the synthesis of knowledge and will. Do we have to recognise these modes of mental function as being separate and in opposition to each other, or do they unite in reaching some synthetic and complete rendering of reality? Kant's resort to the noumenal, as opposed to the phenomenal, only sharpens the distinction between theoretical and practical intuition by justifying the trans-subjective postulate of the one, while denying the trans-subjective presupposition of the other. The result so far is a form of voluntarism based on moral intuition.

The further thought of Kant[1] carries him over to a

[1] Contained in his *Critique of Judgment.*

form of absolutism in which the æsthetic judgment plays an important part. This is of importance in the development of theories of the synthetic and reconciling type, as well as in that of absolute idealism. Its consideration in the former aspect is deferred to its proper place below.[1]

22. By way of criticism, we may say that Kant himself demonstrates the futility of the distinction between the phenomenal and the noumenal. The attempt to give transcendental value to the intuitions meets with distinct failure. To get a reality having any intelligible content, resort must be had after all to experience, to the phenomenal, and to the acceptance of its trans-subjective reference. How to hold together, then, the modes of reality discovered by mediating experiential processes and those implied in intuitive immediacy—this remains the question.

Philosophy in Germany after Kant showed a series of efforts properly described as romantic and epic; efforts to construct the absolute on the basis of noumenal reality; efforts which vibrate between the claims of thought and those of will, between intellectualism and voluntarism.[2]

§ 4. *Results*

23. The outcome of our examination of theories based upon the intuitive type of immediacy is on the whole as follows:

The doctrine of intuition, developing into dogmatic and absolutist theories, does not justify our looking to

[1] See chap. xi., § 4.

[2] Aspects of this absolute idealism are presented in other connections below: the treatment of the absolute as super-personal (Bradley), the resort to the æsthetic (Schelling), etc.

immediacy of this type for an exclusive or final revelation of reality. For this sort of immediacy is empty, apart from the filling it receives from concrete experience of the mediate type—that is, from experience of actual knowledge and will—much as the first immediacy, that of the primitive, is blind except for the same resort. One lacks content, the other form.

This is in itself, moreover, a demonstration of our position that of these two immediacies, the one, found at the beginning of the genetic progression of mind, looks forward toward the mediating processes in which it is to receive its enlightenment; while the other, coming latest in the movement of genesis, looks backward to the same processes for its content or embodiment. The one effects contacts which are unintelligible; the other lacks contacts which are more than formal. The one gives material which is shapeless; the other issues in a hollow shell without material.

The lack in each case is evident. We can not dispense either with the concrete intellectual discoveries or with the direct practical guidance which the exercise of our mediating faculties of knowledge and will alone affords.

CHAPTER XI

IMMEDIATE THEORIES: (II) THOSE BASED ON THE IMMEDIACY OF SYNTHESIS

§ *1*. *The Immediacy of Synthesis*

1. A third type of immediacy appears when processes themselves relatively distinct, and mediate in their type, fall together in a whole of synthesis and apparent reconciliation. The unity of self-consciousness, for example, seems to be in a sense a synthesis of many processes present together; the unity of volition, the synthesis of the motives or active elements which lead up to conscious decision; apperception in general, a synthesis on the side of content which gives a value of immediacy. In each of these cases, the synthetic result, in which the elements lose their separate character or function, is a new state of mind, something *sui generis*. Its rôle in the apprehension of reality is disclosed in the character of its object. As a state of mind, it must be considered as for itself *not composite*, though issuing from the partial motives or factors which have entered into it.[1] We should not read into it meanings which belong genetically either

[1] It is one of the canons of genetic logic that each succeeding psychical mode is to be understood only by its own inherent and actual characters: the canon of "progression" (with the fallacy of "composition"). See *Thought and Things*, vol. i., chap. i., sect. 23.

before or after it in the development of consciousness; but we must take its meaning only for what it is.

Among the theories of reality based on immediacy, an important group make appeal to synthetic states such as these, to illustrate the immediacy intended. Besides the theories of primitive and transcendent immediacy, already spoken of, we find a further group based upon the immediacy of synthesis or reconciliation.

§ 2. *Use of the Synthesis of Personality*

2. Higher Mysticism. The groping for the real, characteristic of early or primitive mysticism, was directed toward objects of mystery, fear, awe, and led to states of absorption, ecstasy, trance, social frenzy, and actual mania. Although vaguely animistic in its earliest forms, it was not directed inward upon the self, but outward upon the world or upon a mystic presence, embodied in the portent, the magical rite, or some other symbol. It was only after the rise of the distinctions in which the subjective was in some degree isolated, as being an inner world, that such mysticism could attempt to justify itself by reflection. This became possible in Greek thought after the subjectivism of the Socratics had prepared the way for the doctrine of conscious personality. In this doctrine, the mystic tradition took on another and more rational phase. It was different from primitive awe and ecstasy, and also from mystic absorption and exhaustion in the transcendent, in that it undertook to justify itself by a more or less conscious use of the notion of personality.

3. This appears in the Alexandrian mystics, Philo and Plotinus. Forsaking causal and logical explanations, they depicted the creation as a progression of the

world from God, in emanations or concentric circles; it was simply a happening, arising in the personality of God and issuing in other personalities. In Plotinus, this becomes an explicit declaration of the presence of the One to the soul, in the state of ecstasy by which the individual returns to God. The world processes are happenings within the all-embracing One, whose first movement is to give birth to personal beings.[1]

Both in the outcome—the abrogation of the dualism between the self and the world—and in the method of apprehension—direct contemplation of simple happenings, taking the place of logical or causal interpretations—it is to be understood as a mysticism based upon the synthesis of personality.[2]

4. The resort to personality was renewed in the mysticism of Boehme at the beginning of the seventeenth century. To him, as to Plotinus, personality was the all-embracing category, reconciling the oppositions of thought and will. But while to Plotinus, the speculative thinker, the absolute was the One, a presupposition of the diversity of personal units and motives; to Boehme, the pious Christian given to meditation, it was a real or ontological first-stage in

[1] In the postulate of mind as pure entity or the One, preceding the movement from which personality issues, Plotinus goes over to a sort of transcendence. But this seems merely to have supplied a logical background to his spiritualistic mysticism.

[2] "By an act of contemplation or direct intuition, the human soul vindicates its oneness with the divine. The will goes out in ecstasy, the heart in love; the will subsides in self-suppression, the heart in a trance-like calm. The divine presence, not revealed to thought or attained by effort, is taken up in feeling by a movement of personal absorption. Here we see the legitimate development of Platonic love, freed from its rational presuppositions."—The author's *History of Psychology*, vol. i., London ed., p. 75.

the development of the divine nature. God is self-generated through successive oppositions and reconciliations in his own nature. The mental faculties, knowledge, will, and love, arise in the course of this evolution of positive personality. There is a constant movement of adjustment—assertion, limitation, reconciliation—in the bosom of personality; and in this the motives of partial process are absorbed and unified.

This idea lent itself directly to the romantic interpretations of later German philosophy, in which the reconciliation of contraries plays so large a part; particularly to that found in the absolutism of Fichte and Hegel. But as soon as it was taken out of the atmosphere of mysticism and theological faith, it became transformed into one or other of the two alternatives of the theories of mediation, intellectualism or voluntarism. Advocates of pure thought, on the one hand, declared, "in the beginning was the word"; while advocates of pure will, on the other hand, retorted, "in the beginning was the act." And so systems of absolutism, based upon alternative interpretations of thought and will, repeated themselves.

5. In this development, history itself puts in evidence the criticism to which theories based upon the synthesis of personality are exposed. The postulate of immediate consciousnes of personality may be utilised in mystic contemplation and theological faith; but as soon as a justification of this is attempted, such personality either falls apart into the fragments of mental analysis or becomes transcendent in an absolute and formal principle, whether thought or will. It is one thing to say that in our sense of personality we are immediately aware of a synthetic unity in which the

various powers of the mind in some manner inhere; this is but to recognise the fact which mysticism realises. But it is another thing to say that the personality which thus apprehends reality is itself the synthetic form of the reality that it apprehends. The dualism of "self and other" asserts itself; and in overcoming this dualism, the Fichtean and Hegelian absolute performs its most startling feats. It "posits," "acknowledges," "asserts"—all substitutes for the old mystic term "creates"—the "other," which must in some manner arise out of the absolute unity and identity of the One.

In fact, these logical interpretations do not avail more than the simple faith of Jacob Boehme. It is perhaps in the mystic moment itself that the objective self does reach real immediacy; but ordinarily this self is a complex of sensational and social elements in which the processes of mediation are more or less evident. To resort to this sort of a self is to return to dogmatic spiritualism, with all its difficulties, and abandon the high road of the immediate. As has been remarked already, Boehme himself indicated this alternative in finding opposition and distinction inherent in the self-generating nature of God.

The difficulty of finding any real and concrete difference between mind and body, on the basis of a self-identical absolute principle, was the motive to Spinoza's doctrine of "substance and attributes." In the attributes, we recognise the duality of concrete existences; in substance, the identity of the absolute. The attributes are defined as "thought" and "extension"; the substance is expressly declared to be undefinable. Nevertheless it is called God; and in this we see the motives of religious mysticism reasserting themselves. In spite of the profession of a "geometrical" method, in Spinoza the antinomy between ideality and actuality, the infinite and the particular, seeks an emotional solution.

6. The Super-personal. In religious thought, as we have seen, the idea of personality is carried over to the absolute which takes the name of God. He is the Super-personal, realised by faith but not adequately apprehended by reason.

In contemporary philosophical discussion, such a super-personal absolute has been approached in an interesting way by those who seek to preserve the point of view of subjective personality, while laying stress on the immediacy of feeling or "sentience," together with the realising of which feeling is capable. The super-personal is made a sort of limiting case, as if reached by means of a series of approximations in the expansion of the self. The various mental powers are conceived as carried on to infinity. This is to suppose a larger all-but-infinite personality, or an "absolute" whose personality is ambiguous, present in each individual "finite centre."[1]

There is, however, no further justification for considering the self as synthetic principle of reality, beyond the identification of the organisation of the mental life, with its show of "unity in variety" and "identity in difference," with that of the world in general. But, as we have seen, it is in this organisation—and when we remain true to it—that our interpretation finds itself embarrassed by the sharpest dualisms and oppositions: those of mind and body, fact and ideal, truth and value.[2]

[1] *Cf.* Appendix C. on the position of F. H. Bradley.

[2] What our personality suggests itself as being is not a whole of all the elements of reality, but part of a larger whole, physical and social.

The projection of the entire picture of the mental life upon the larger cosmic canvas does not seem to rid it of its inherent disharmonies.

While such formulas as unity in variety, identity in difference, and organic relation of whole and parts, do present reassuring figures drawn from the more formal aspects of the life of the self, they do not really give us light on the material nature of the world.[1] At any rate we may say—recognising the profundity of these attempts to interpret the conflicting motives of the real[2]—that while these aspects of conscious organisation do go some way to illustrate the nature of reality, other modes of self-conscious unity may be pointed out which take us further still. Instead of simply assuming by analogy that the nature of reality, in its completed organisation, repeats or expresses the unity of organised experience, we may find evidence of this identity, where Kant thought it possible to find it, within mental function itself. Is there not, we may ask, some mode of experience in which the self realises its true nature by an identification of itself with its own organised content? To this question, we are able in what follows to give an affirmative answer.

7. But the mere expansion of personality accomplishes nothing. As the factors involved are one by one pressed on to infinity or made to fulfil their ideals, each in turn loses its distinctive character in the unity of the whole; and in the end the colourless white light of formal identity succeeds to the coloured rays of the spectrum of personal experience. As an abstraction,

[1] A discussion between Prof. B. Bosanquet and the present writer, on this and related points, may be referred to in the *Psychological Review*, 1902–03.

[2] The position has been worked out very brilliantly by the school of British neo-Hegelians, Bosanquet, Bradley, Creighton, and others.

the super-personal loses both the concrete immediacy of the mystic's feeling and the qualitative concreteness of the knowledge and will postulated by intellectualists and voluntarists.

8. It remains true, however, that in the resort to personality, considered as a synthesis in which opposing functional elements are united, the requisites of a solution of the problem of reality are better realised than in any other of the theories mentioned. Its limitation is, that just in the full realisation of the self, the opposition between the person and the thing, and also that between distinct persons, is made most acute. There would seem to be no way to obviate this difficulty—apart from the resort to religious mysticism—except by a re-examination of the entire movement of consciousness, with a view to detecting some further reconciling motive. As we have intimated above, the question is this: is there any experience in which *the self realises itself, not as in opposition to the "other," but as in the other?* Such a type of experience, projected on the broader screen of the world, would seem to be what the various solutions—rationalist, voluntarist, affectivist, each isolating a partial motive—have alike sought for. It would also issue in a philosophy in which personality, understood in the broad sense of "experience," would be fundamental.

The way to such a point of view has been prepared for, in my opinion, by the attempts to develop the theory of feeling on the epistemological side. This theory, fragmentary as it has remained, and sporadic in its recurrence, nevertheless promises to take on the form of an articulated and intelligible doctrine.

§3. *The Synthesis of Feeling*

9. Platonic Love.[1] In the "divine love" of Plato, the earliest and in many respects still the most interesting attempt to reach a synthesis of this kind was made. Plato develops his theories of the good and the idea and finds that these both alike lead to the highest idea or God. God, the supreme idea, is also the supreme good. It is reserved for feeling, however, to realise the idea of God, as being both absolute reason or idea and absolute end or good, in a single intuition. This takes the form of the immediate contemplation of the divine, the state called "divine love." The mediation of the real through ideas and that of the good through ends, both intrinsic elements in Plato's philosophy, issue in *the realisation of God in immediate feeling.*

Plato is not, however, simply to be classed with the religious mystics. His love is indeed religious, because the supreme idea is God; and it is also mystical, because it is a state of feeling which of itself gives no account of its object. But it is none the less a synthetic result, in which the elements of reflection pass into a higher immediacy; it is not a mere lapse into feeling, after the abandonment of thought.

This last statement is reinforced in view of the æsthetic factor in the Platonic love. This love takes the form of an immediate dwelling upon the perfect and ineffable, a rapt contemplation in which perfection is realised and immortality attained. In it there is the response of the divine reason in man to the divine goodness and wisdom of God. In thus realising the

[1] On this and the succeeding headings of this chapter, the treatment of W. D. Furry (*The Æsthetic Experience*, the historical part, especially chap. vi.) will be found suggestive.

absolute idea, the contemplator grasps the æsthetic ideal; for this ideal, that which the artist imitates and that of which he creates the semblance in the work of art, is for Plato the archtype or idea of the actual thing of nature.[1]

The tradition of Platonic love divided itself, in the historic event, into two currents. The religious motive, involving the personality of God, passed into the mysticism of Plotinus and Boehme, as we have already seen; and the philosophical motive took on a more psychological phase in the attempt to locate the function of feeling and appreciate its epistemological value. This latter is the beginning of what we may call Affectivism.

10. Affectivism. In its simplest forms, this was simply an attempt to isolate the psychological basis of mysticism and so to justify it. In the German mystics, Eckhart and Tauler, at the beginning of the fourteenth century, the heart—so to translate the German term Gemüth as used by them—the affections, the entire conative-affective determination of personality, is involved. The heart, as over against the reason, is the complete revealer of reality; it takes up and completes the work of thought. It is a synthetic function; one of immediate realisation, not one of analytic comprehension.

[1] *Cf.* Gomperz, *The Greek Thinkers*, vol. iii., pp. 102 ff. In the *Symposium* of Plato, we read: "He who has been thus instructed in the things of love . . . will suddenly perceive a nature of wondrous beauty . . . absolute beauty, separate, simple, and everlasting, which without diminution and without increase, or any change, is imparted to the ever-growing and perishing beauties of all other things. . . . This, my dear Socrates, is that life above all others which man should live, in the contemplation of beauty absolute. . . . Remember how in that communion, only beholding beauty with the eye of the mind, he will be able to bring forth not images of beauty but realities, for he has not hold of an image, but of reality." As translated in Bakewell's *Source-Book in Greek Philosophy*, pp. 156 f.

11. Constructive Affectivism. The function of apprehension that appears in the movements of the heart and in "illumination," in general, was further defined by the mystics of the Renaissance in the sixteenth century. They developed the Aristotelian doctrine of the imagination, which was looked upon as mediating, in various ways, between sensation and reason.[1] Imagination completes the detached data of sense, building them up into ideas, and offers preliminary schemata or ideal constructions to the reason. This is an anticipation of Kant's view of the schematising imagination and on the whole a clearer statement of it. It also suggests the modern doctrine of the assumptive or experimental function of the imagination.[2]

It is interesting that this should have been hit upon by writers of a mystic cast of thought. It constitutes an important step in the development of mysticism out of the status of emotion and sentiment into that of a rational constructive theory. If the imagination in its normal working accomplishes the results formerly attributed to emotional intuition and ecstasy, then this type of apprehension may be put down as one of the recognised functions of cognition.

12. A synthetic working out of the topic is not to be found in these mystical writers, but their view should be taken account of in any theory, such as our own, which finds in the imagination the intellectual skeleton about which the elements of affective synthesis take on their higher form. It is in the schematising and prospecting imagination that mental constructions of

[1] *Cf.* on this passage the writer's *History of Psychology*, Lond. ed., vol. i., pp. 101 f.

[2] The application of this view to the "semblant" products of play and art is spoken of later on.

all sorts—knowledges, ideas, ends—are released in some measure from the strict control of fact and utility, and rendered available for the more direct realisations in which the self unites itself with its object. Without this reconstruction of the object in semblant form by the imagination, no intrinsic reconciliation of the objective and subjective would be possible. It is just in this respect, indeed, that the theories already examined alike fail. They either subordinate one of the opposing terms to the other, or conceal them both beneath a covering of feeling or intuition. An essential reconciliation demands a function in which the root of the opposition between subject and object is removed, and the dualism of inner and outer controls *cured at the source*. That function, we are convinced, is the imagination.

13. Faith. The next step toward the rational theory of the affective is the "faith philosophy," especially as we find it in the writings of F. H. Jacobi. In Jacobi, however, it is not the intellectual framework that is under investigation, but the attitude of mind and the method of procedure found in immediate apprehension. This attitude, he declares, is one of faith, which differs both from the rational conviction produced by proof, and from the credulity of childish trust. It is the attitude of immediate acceptance by the mind, when in the presence of reality, sensible and rational alike.[1] And the appropriateness of the object

[1] Jacobi's view recalls to mind the later theory of Brentano who makes the function of judgment an original active attitude of acceptance or rejection. But to Jacobi faith, although it may be excited in response to argumentation, is not limited to this; it is more universal and also less fallacious than intellectual conviction.

Another position interesting to recall here is the theory of "second thought" of the Franciscan, Duns Scotus, who held that a suggestion

to excite this attitude becomes the criterion of reality. The real is that which manifests itself in a way to excite our faith.

14. This is a step away from mystical ecstasy toward the theory of intuition; and indeed Jacobi identified faith with the pure reason of Kant, finding it to be, however, a state of immediate feeling. As a further determination of immediacy on the subjective side, it is an advance; but in its characterisation as a state of faith, it does not reveal the motives of a genuine affective synthesis. It should be supplemented, on the cognitive side, by some motive which would render objectively valid the objects of faith. Taken simply as such, "fideism"—as the faith philosophy has been called—is more subjective, on the formal side, than intuitionism which developed from it; but it is little less blind, on the material side, than the mysticism in which it had its origin. It needs just the doctrine of the imagination, or some equivalent, to supply an intellectual framework: some process of construction of the true and the good in a form which faith may utilise without coming into opposition with actual fact or unfulfilled desire. The reconciling immediacy, we may say again, must be one of real synthesis of cognitive and active factors.

§ 4. *The Æsthetic Synthesis*

15. The synthetic reconciliation we are seeking has been found, more or less incidentally but none the less really, in the æsthetic experience by three of the great thinkers of history. Each of them represents a move-

coming into the mind as a "first thought," had to be actively embraced or accepted by the will, to become a "second thought," to which reality and the agent's responsibility attached.

ment of thought at its climax—at a time when all the difficulties of alternative logical hypotheses had become evident. They are Aristotle, Kant, and Schelling.

(1) In Aristotle, the motives of earlier thought met and mingled. The subjectivism of the Socratic appeal to the inner world and the objectivism of the scientific and cosmic points of view required synthesis and reconciliation. In Plato's thought, the absolute idea and the absolute good each demanded place in the real, in God; and in Platonic love an affective mode of reconciliation appeared—having important æsthetic value as we have seen—which came to its culmination in the mysticism of Alexandria.

In Aristotle, this reconciliation passed from the status of feeling to that of theory. The distinctions between potentiality and actuality and between matter and form, required that God be pure form, pure actuality. As such, his existence or being is one of eternal self-contemplation; he is apart from the worlds, which are held to him and to their proper place and movement by the spiritual principle in them. They are moved by love to God. In this cosmic theory, a sort of concrete illustration is given of the love by which the spiritual part of man seizes upon the reality of God. In return, God dwells upon the universe as upon a work of art, a completed whole in which the ideal of the æsthetic imagination is presented in sensible form.

16. It would not be difficult, of course, to exaggerate the theoretical importance of this first synthesis; it is still semi-mystical and sufficiently figurative. It mingles love with law, spiritual with cosmic forces. But when we remember that Aristotle's physics was teleological in principle, not mechanical, and that the end or *telos* was pure form, quite free from matter, we see that it

is not so lacking in relevancy as might at first appear. It was also not to go far astray from his metaphysical principle for Aristotle to find in the artist's ideal, and in its apprehension through contemplation, the point at which the teleological movement of things comes to its point of repose in the completed picture present to the contemplation of God. If there is an absolute intelligence or experience, and if the sensible world is objectively apprehended by it, does not æsthetic contemplation afford the best analogy by which to represent the relation of the one to the other?

17. (2) To Kant, another of the three thinkers mentioned, equally urgent alternatives were presented. Logical rationalism had developed into dogmatism and formalism; sensational empiricism, into scepticism and materialism; spiritualism into intuitionism and religious mysticism. In Kant, these philosophical oppositions repeated themselves in the form of closer distinctions between reason and sense, the pure and the practical, form and matter. Theoretical reason lost itself in a maze of antinomies and did not reach reality. Practical life assumed or postulated the real, but gave no adequate reason for its confidence. The inference was plain that life without insight can not achieve its full end, and that reason without content can not people the dwellings of life.

Is there, it may be asked, any more intrinsic bond between the true and the good, between the theoretical and the practical reason, between the "nature" which intelligence presupposes and the "freedom" which morality assumes? Is there any bond between the formal or *a priori* as such which the reason legislates, and the concrete facts and motives of life which sensible experience contains?

In the *Critique of Judgment*, Kant finds, or at least intimates, a mode of reconciliation of thought and nature, or as he puts it of "nature and freedom," in the domain of feeling, in the judgment of æsthetic taste.

18. The distinctive object of the judgment of taste to Kant is not the objective content itself; this owes its form to rational and practical judgments. In the judgment of taste, on the contrary, we have the *a priori* principle of the purposiveness, suitableness, or fitness of the world for apprehension by the mind (in the case of the judgment of the beautiful), and of the mind for the apprehension of the world (in the judgment of the sublime). But while æsthetic contemplation remains formal and regulative in its principle of judgment,[1] still in the accompanying experience of æsthetic pleasure and its contrary, we have immediate concrete intimations of the larger relations of harmony as between the mind and its objects, and as among the faculties themselves.[2] Thus the place of mind in the whole of reality and the meaning of reality for the mind are in a measure disclosed by being translated into feeling.[3]

19. An important adjunct to such a synthesis, if it is to be effected by means of the æsthetic experience,

[1] *Cf.* Basch, *Revue Philosophique*, Jan. 1912, pp. 23 f.

[2] "The æsthetic predicates (to Kant), at all events the beautiful and the reverse, are to be explained as consequences of the greater or less ease with which the particular and the universal are united in the concrete apprehension of the thing" (R. Adamson, *The Development of Modern Philosophy*, vol. i., pp. 226–7). The universal is the subjective, the particular the objective factor.

[3] It is here that Kant reaches, but does not develop, the postulate of a monistic principle, by which the distinction between the soul and the world, freedom and nature, is abolished. He still pursues his quest for the "thing in itself." "There must be," says he, "a ground for the unity of the supersensible, which lies at the basis of nature, with that which the concept of freedom practically contains." (*Crit. of Judgment*, Bernard's trans., Introd., p. 12.)

is the theory of the imagination. We have noted the use made of the imagination in the theory of the æsthetic by Aristotle, and the extension of its cognitive rôle by the Italian mystics. In Kant, we find a doctrine of "schematism," in which a part is assigned to the imagination not only in the formation of special productions such as those of art, but of all the normal products of knowledge. The schematic imagination, arising between perception and thought, is a mediating construction. It throws the manifold of sense, as by a first synthesis, into schemata or programmes for the further work of the intelligence. It is a sort of functional link, making easier the passage from sense to reason, and lessening the formality of the intellectual categories.

20. Taken together, these positions seem to afford materials for a synthesis of will and reason in the domain of æsthetic appreciation. Kant intimated the nature of such a synthesis, as residing in the constructive rôle of the æsthetic imagination; but if he had carried out his idea consistently, it would have taken the form of a synthesis in the domain of formal ideals, effected by the judgment of taste, which reports *a priori* the harmony of thought and things.[1] Feelings of pleasure are only incidental to these formal relations.

It is plain, in any case, that Kant did not fully utilise his doctrine of the schematic imagination. The concrete

[1] Too much importance should not be attached, indeed, to the æsthetic intimation of such a ground of unity or synthesis, in Kant; since for him there are others equal or superior to it, notably that found in the purposiveness shown in the organisation of living creatures. Such intimations are empirical points of view taken up for purposes of interpretation rather than intrinsic features of "criticism" as such. I am inclined now to think that in another place (*History of Psychology*, Lond. ed., vol. ii., pp. 24 f.) I attributed too definitely to Kant the intent of an explicitly æsthetic reconciliation of pure and practical reason.

schema of imagination was for him merely a phenomenal product, a step in the ladder of empirical functions. It did not suggest itself to him that it was one in nature and rôle with that judgment of taste by which the synthesis of ideals was effected. This would have broken down the distinction between the formal and the empirical.

Instead, therefore, of overcoming the opposition between the phenomenal and the noumenal, between sense and reason, between will and intelligence, it presents, at the best, a new and formally more synthetic ideal of reason, that of taste. The objective value of this ideal is still open to all the ambiguities of the distinction of the phenomenal and the real. Granted that in ideal beauty one may discern a pattern of ideal truth and of ideal goodness existing in harmonious unity, how far, it may then be asked, does this secure truth and goodness in the elements of the actual works of art we admire. And if it does, what is the mental function by which this synthesis is achieved? In other words, is there an empirical synthesis, a real union of cognitive and active factors, in the æsthetic experience itself? This Kant might have found to be the case if he had proceeded out from the schematic function of the imagination as he himself recognised that function.

It is this defect of undue formalism that further theory should attempt to remedy. Instead of starting out with the formal and universal, we should start out with the empirical and concrete. The synthesis we seek is not one to be welcomed as fulfilling the exigencies of logical and formal construction, but one to be recognised as bringing to light, in actual experience, a motive of reconciliation.

21. The movement in the direction of a constructive

affectivism, in the sense of pancalism, as developed below, received in Kant, however, when all is said, a powerful impulse. He suggested reasonable ground for finding the union of will and reason in feeling, and for finding this feeling, as had been proclaimed by Platonic mystics and Aristotelian absolutists alike, present in the exercise of the æsthetic judgment. Moreover, he confirmed the rôle of the imagination, as supplying to art its cognitive framework—a position also suggested by earlier thinkers. What is still lacking, however, is the integration of this entire imaginative mode of synthesis in the body of concrete experience; so that we may say that it is possible always, in the life of action and thought, to rest upon and enjoy —to realise, in short—a beautiful reality. This task is to be accomplished by psychological research and analysis, rather than, as in Kant, by the critical selection of one among several formal alternatives.

22. (3) A third thinker, Schelling, found in æsthetic contemplation a synthesis of the factors recognised as urgent in the philosophy of his time. The romantic assertion of the self, by Fichte, in the form of absolute self-consciousness, had reduced nature to a negation; nature was merely the limitation posited by the ego in the process of its development. To Schelling this failed to do justice to the reality of nature. Rather, said he, nature is an earlier stage of the ego itself: its unconscious, prehistoric stage. The same motives of evolution work in the unconscious forces of matter, until they break through in the consciousness of organised life.

The dualism, then, between mind and matter, conscious and unconscious, was to Schelling one of evolution; mind slumbers in implicit form in nature. And the question arises—is there any mode of apprehension in which their real identity of essence shows itself? Does conscious life have any way of testifying to the absolute identity of the spiritual principle, and to the unity of all reality?

To this question Schelling replies by pointing out three functions or activities of which conscious spirit shows itself possessed: the activities of knowledge, practice, and art. In the last named of these, the spirit attains the intuition by which it realises its identity with nature. In art, the conscious claims the unconscious as one with itself.

23. Besides this synthesis of mind and nature in art, by which the philosophical principle of identity is justified, Schelling finds in art also another essential synthesis, that of knowledge and practice. In æsthetic enjoyment, a faculty of heavenly fancy, a spiritual intuition breaks down the opposition between the theoretical and the practical, and brings the spirit into consciousness of its unity with itself.[1]

[1] Of these two syntheses, that of the identity of nature and mind, revealing itself objectively by a genetic or evolution process, remained in German speculation. But the latter, the synthesis of theoretical and practical in art, was not influential. On the contrary, the contrasted motives of thought and will worked themselves out respectively in the opposing systems of Hegel's intellectualism and Schopenhauer's voluntarism.

The resort to art made by this latter writer, Arthur Schopenhauer, it is interesting to note, is of a different kind. To him the æsthetic is not a synthetic experience, but a temporary detachment, by which one gets a momentary deliverance from the torments of a life of unsuccessful struggle. The will is for the moment suppressed. Art is a retreat, a luxury. Schopenhauer overlooks the element of real life in art, the inclusiveness of art. "He forgets," says Höffding (*History of Modern Philosophy*, Eng. trans., ii., p. 234), "the sympathetic absorption which indeed presupposes that we attribute worth to the æsthetic object." The real deliverance is found by Schopenhauer in the state of quiescence of the ecstatics or mystics, in which the will-to-live of the individual is lost in the "nothingness of Nirvana."

The following passage from Adamson states clearly the spirit of Schelling's teaching.

"Theoretical consciousness terminates in that abstract attitude of inner reflection in which self seems to be simply and absolutely opposed to all concrete content. From this position escape, reconciliation of difficulty, is found in practical activity, in the realisation of self in action or conduct. But practical consciousness, nevertheless, even in its highest development, still leaves unreconciled the thinking subject, conscious of himself, and the sphere within which his activity lies; or, put otherwise, the highest phase of practical consciousness brings sharply before us the opposition between the theoretical and the practical, that opposition which in the Kantian system is formulated in the distinction between the realm of nature and the realm of morality. Fixed, necessary, universal—these are the predicates as far as theoretical comprehension of fact is concerned: spontaneous, free, individualising—these are the predicates we assign to the practical self. Reconciliation of the opposition—a conception which shall give due recognition both to the universality and fixity of nature as known, and to the spontaneity, freedom, of spirit as realising itself in action—is given in the representation of nature as a kingdom of ends, and in the attitude of consciousness thereto which finds expression in art. The artistic view of nature wherein reality is taken as a living whole, the expression throughout of spirit—this, for Schelling, is the highest reach of thought, the final attitude of speculation."

And again:

"Doctrine of knowledge, doctrine of morality, doctrine of art, these are the three divisions of the philosophy of mind; and the progress is, in each of them, of the same general type. The advance, as it were, from each stage is effected by the appearance of differences or problems which call for a higher, richer, more comprehensive mode of consciousness.[1]"

24. These positions of Schelling's have the great merit of showing the essential need of synthesis, and of pointing out the factors which must be included in it; he also finds the positive mode of consciousness in which the reconciliation occurs to be the æsthetic, the art consciousness. The great dualisms of genetic process—mind and matter, knowledge and practice—these are, as we have found, the chronic oppositions of the mental life; the thorn in the side of a rational theory of things.

But the solution offered by Schelling, while in our opinion pointed in the right direction, is too speculative and also too rhapsodical to be convincing. It vibrates between the abstract indefinite and the mystic immediate. The art consciousness is not shown to have the requisite synthetic content.

The persistence of the dualistic alternatives, and their refusal to down at the waving of the æsthetic wand, is shown in this philosopher's subsequent philosophical thought.[2] In his very latest work he concludes that speculative formulation of the nature of reality must remain merely formal; and that the essence of the real, its content, is apprehended only in direct experience of life and in the testimony concerning that life found in history.

[1] R. Adamson, *The Development of Modern Philosophy*, vol. i., pp. 266–7.

[2] In Schelling's second period, seeking to state the principle of identity which the æsthetic consciousness reveals, he reached an undefinable Spinozistic absolute; and then, in his last period, aiming at making this concrete, his thought swings over to mystical expressions borrowed from Bruno and Boehme.

What Schelling's resort to the æsthetic really lacked, then, was an analysis of the art consciousness and its products, which would show that it really fulfilled the rôle of reconciliation and synthesis which he assigned to it.[1]

[1] Our exposition confines itself to these historical thinkers; but other important suggestions have also been made as to the epistemological and synthetic value of the æsthetic, such as those of the poet Schiller and Lotze.

Of the recent attempts to deal with the problem, that of A. T. Ormond (*Foundations of Knowledge*, 1900, pp. 227 ff.) is the most interesting. See Appendix B.

CHAPTER XII

RESULTS OF THE HISTORICAL SURVEY: THE DEMAND FOR AN INTRINSIC SYNTHESIS

§ *1. The Presupposition of Truth and the Postulate of Value*

1. Our brief exposition of theories has thrown into relief the alternatives of speculative thought, and also its shortcomings. As to the alternatives, we find that they exhaust the possibilities of implication and postulation. For understanding these terms about in their usual meaning in the discussions of logic,[1] we now see that the final problem lies in the interpretation of the relation between these two. What is the ground of correlation, if there is any such ground, between the realities implicated or presupposed by the faculties of cognition on the one hand, and those supposed or postulated by the faculties of appreciation, valuation, will, on the other hand?

One of these sorts of reality, that of implication, is called actuality; it resides in things taken as existing in a cosmological or ontological order, in a sense finished, and constituting a completed whole. It presupposes a sphere of existence of some sort—fact, truth, relational implication of part in a whole—a sphere set up as

[1] Cf. *Thought and Things*, vol. ii., chaps. v, x.

fundamentally characteristic of reality. Reality must exist; it must be actual, in order to be real.

The other type of reality is reached in the play of the active and appreciative life. To it the final form of reality is that of something taken to be satisfying or good, approved of the faculties of appreciation, desire, and will—in a broad sense, ideal. The existing or actual, in order to be real, must also be valuable. But the final form of value is not reached as actual or as a predicate of existence; it is supposed, assumed, set up as a demand, a model, an ideal; it is a "postulate," not a presupposition or an implication.

2. Our historical survey has shown the rivalry of theories of these two types: on the one side, theories ranging from materialism to refined subjectivism, characterised by the presupposition of actuality. Ideality for them is derived, secondary, consequential. On the other side, there are the theories of ideality, finding in the actual merely the sign, symbol, *locus*, of the values with which the life of will and sentiment is charged. Here the theories range from pragmatism to absolute voluntarism and idealism of value.

Later philosophy has lent itself to the task of reconciliation. Its constructive effort is to bring together the presupposition of actuality and the postulate of value. On the whole, it has not succeeded; for in each case it has had to extend the range of one of these terms into the field of the other.

On the one hand, knowledge, we are told, is after all motived by will and has its ideal in a value: hence the justification of voluntarism. But granted the point, still does not the real need a point of actuality in which its values may reside?—can the ideal float in a void without presupposing an existing real?—and how can we get this but through processes of knowledge?—how establish values except by establishing facts and truths?

So in turn intellectualist theories "cut under" the postulate of the voluntarist. The presupposition of existence, it is said, underlies the meaning of value; the only real value is that which is achieved, attained—in short, that which is found in the realm of existence or actuality. Ideal values are chimerical; "that which a man hath, why doth he yet hope for?"

To this the voluntarist replies: "Yes, but why then do you pursue the ideal you call truth? Why do you join in the chase to which all life and all creation lends itself—the chase after ideals? Is the satisfaction of striving and the joy of aspiring, worshipping, contemplating, to be entirely left out of our account of things? Are the requirements of will and feeling, the postulates of God, freedom, morality, to be simply neglected when we attempt to give an account of the whole nature of things?"

3. In the Kantian *Critique*, a great attempt at a morphology of reality, no permanent adjustment of theoretical and practical is reached. The practical postulate is given precedence as guaranteeing its object; but the distinction between phenomenal and noumenal gives the case in advance to the intellectualist; for it makes a division in the modes of actuality and reserves reality for that contained in reason as opposed to sense. The entire question for Kant becomes this: how is reality according to reason possible? The answer is, perforce, a sort of formalism.

The Post-Kantian theories vibrate again between intellectualism and voluntarism, under various names, the whole controversy becoming, however, after Kant, more transcendent. This has motived the reaction to experiential and pragmatic theories, to mysticism, and to the solutions to be found in a synthesis based on feeling.

4. This rivalry between the presupposition and the postulate—between the rational implication and the desirable ideal—shows itself also in another form. The theories of actuality are not content to accept the various modes of the actual which experience reveals; this would be too pluralistic. All actuality must be reduced to one type: it must be perceptual, or rational, or moral, or spiritual. So the actualist shifts his ground, proposing an ideal of unity for reality, and introducing a monistic postulate in order to overcome the dualisms and pluralisms of his own requirement of actuality. The great dualism of mind and matter is in this way overcome when the presupposition of mind is carried over as a postulate into the domain of matter, as in spiritualism and theism. Materialism, on the other hand, extends the sphere of matter, in the form of brain cells or chemical elements, in order to explain the reality of mind. Its presupposition is also converted into a monistic postulate.

So through all the turns of systems of monism, an implication or presupposition, quite legitimate in its own sphere of control, is erected as a postulate in another sphere which stands in contrast or opposition to the first.[1] So one has, in the mosaic of philosophy, as many monisms as one could desire.[2]

5. Our historical sketch has also brought out the

[1] A procedure that is in direct violation of a canon of genetic logic. See *Thought and Things*, vol. i., chap. i., sect. 26, v., the "Canon of Modal Unity."

[2] On pluralism and the grounds of monism, see below, chap. xiv., § 2.

fact that a reaction tends to show itself, whenever speculation has reached its full maturity, toward the recognition of experiences of immediacy, experiences in which the dualisms and oppositions of actual and ideal, the presupposed and the postulated, have not been developed, or in which the motives which produce these oppositions spend themselves in states of fulfilment or exhaustion. Hence appear various attempts to find an immediacy, whether of revelation or of intuition or of inspiration or of contemplation, in which a purer and fairer vision of the real emerges upon the gaze of man.

The really synthetic approaches to affectivism have not gone far; but the beginnings are significant, as we have seen. In attempting to carry this line of thought further, we are simply interpreting the results of detailed investigations, which point plainly in this direction. Before proceeding further, however, we should say a word in defence of the method we employ, since it seems to reduce the high ontological problems of abstruse speculation to the level of the modest undertakings of genetic and empirical science.

§ 2. *The fruitful Method, genetic*

6. In the earlier work already cited, we have distinguished certain points of view from which the activities of knowledge may be considered.[1] The three points of view respectively of the "logician's logic," the "metaphysician's logic," and the "knower's logic" were described and compared; and the point of view of the knower himself was adopted and defended. Reasons are there given for the adoption of this standpoint, together with the genetic method to which it lends itself;

[1] *Thought and Things*, vol. i., chap. i., § 1 f.

but we have now arrived at a point at which a more complete justification, both of standpoint and method, is possible. If we were right in claiming that the processes of knowledge, by which realities are discovered, and the indications by which belief in them is justified, can be traced out and set in evidence relatively to one another, then when this is done in detail, we should be able to use the results in more than a defensive and apologetic way.

In fact, we can now, I think, directly refute both the logical and the metaphysical methods of approaching the problem of reality, in favour of a psychological and genetic method.

7. The logical method, in distinction from the ontological or metaphysical, consists essentially in giving to logical principles an absolute and unconditional value, apart from the material of knowledge to which they have application. Universality and necessity are said to attach to them, being marks of their validity, in the very nature of things. Reality therefore, must be logical in character—it must be constituted by these principles or reached through them. Whatever further sorts of realisation or determination we may find in experience or in nature, these are to be looked upon as being at best secondary and ancillary to the logical. The logical mode of dependence and determination is final and fundamental.

8. Two principal criticisms of this procedure may be made, both justified by the results of genetic research.

(1) In the first place, it has been shown that the universality and necessity of logical rules and laws are marks and characters which arise in the growth of experience of certain sorts. Like all other psychical characters, they have their genetic history, and repre-

sent the culmination of long-continued functional processes. While we would not contend that their genesis prejudiced their validity, or explained fully their force—this would be to illustrate the fallacy of "composition," too often committed by geneticists[1]—still we claim that the fact of this natural genesis forbids our taking these principles out of their context and treating them as having miraculous and mysterious ontological virtue. What they are worth is shown only in what they accomplish; it appears in the rôle they fill in the development of the whole meaning of the real.

As matter of fact, just in the marks by which logical principles show themselves to be universal and necessary, they become in actual life practically useless. We do not argue by them in any strictness; we violate them whenever they come into conflict with other imperatives of a different order. Even science admits their tentative value and hypothetical force. So far from dictating to us the realities we are to accept, they seem, when stated to us, either to come from a distant and unfamiliar world, or to be of the commonplace force of the majestic formula of identity, "A = A." Our intimate realisations occur in the realm of the concrete, singular, immediate, pungent experiences of feeling and action; and in the sphere of knowledge we leave out perforce, in our generalisation and abstraction, those marks of direct apprehension by which the real is made most fully available. Logical thought does not destroy or invalidate the object; on the contrary, it carries further our valid apprehension: but, by reason of its method of employing universals, it serves us in a very partial and inadequate manner.

[1] See *Thought and Things*, vol. i., chap. i., sect. 23. There may be more in the full meaning of a psychical content than the conditions of its genesis express.

Universality and necessity, so far from being presuppositions of reality in all its aspects, are merely ideals arising in the operation of partial and restricted processes of cognition. To read them in advance as prototypes of the real, or as its constitutive principles, is to content ourselves with the chaff of empty form, when we might have the wheat of the real life.

9. The second point is this: the same procedure in kind may lead, and often has led, to the similar reification of the practical universals, the moral imperatives, at the expense of the theoretical. Moralist and voluntarist have as much right to hypostatise their principles as the logician has. But the realities they reach by this method are in many ways opposed to those of logic. They involve the postulate of personal freedom and absolute good, present in the single experience and realised in the flux of feeling and action. Who is to decide between these rival claimants? We may decide—we do so every day—not between them, but both for and against them both, on occasion. The practical norms, like the theoretical rules, are ideals appearing in the course of the working out of concrete psychological and social motives. They have their function here and here only. Practical rules compose the valuable but fallible guide of life just as logical laws compose the valuable but tentative directory of thought.

10. The alternative to accepting this simple result and giving up the pretensions of formal absoluteness of either kind, is the metaphysician's choice: he plunges into the flood of ontology.[1] In view of the balanced sceptical outcome of Kant's *Critique*, which reacts to justify a frank empiricism, the post-Kantians postulate an absolute thought or will, whose logic or dialectic

[1] See the remarks on Bradley's views, in Appendix C.

reveals the real. The question comes to mind whether philosophy is to continue to thrash out the straw of these different but similar absolutisms?

A negative answer to this question is embodied in the various instrumentalisms, intuitionisms, affectivisms, pragmatisms, and mysticisms of the present day. Our own negative, now to be stated, is based upon the results of empirical research; it denies the validity of the ontological point of view altogether.

While cognitive process claims to establish the real apart from thought—an ontological real—this reality has always the meaning given to it in the development of experience. The trans-subjective reference, if accepted, justifies a separate reality; but one whose characters are known and confirmed in the special sphere of control to which the reference is made. The reality of a second person to the knower, for example, is not the same as that of a physical thing; each has reality of a circumscribed and relative sort. So while consciousness seems spontaneously to disown her own children, by finding them to be foreign and independent of herself, still they remain her children.

This has appeared, in the realm of reflection since Descartes, as the presupposition of all modern thought. The subjectivistic standpoint is persistent. It also appears in those modes of consciousness in which the trans-subjective reference is, in certain circumstances, abrogated, even in the cases of objects to which it ordinarily attaches. The same content—say a building—may be an object of imagination, of perception, of memory, of æsthetic contemplation. Now which of these is to stand as the final mode of its reality?—which of them is to be considered the ontologically real? Why should the mode of perception, or the mode of thought, take precedence over the others? For practical purposes, perception is no doubt the most useful; but for æsthetic purposes it is less interesting; and for social and emotional purposes, it may be quite unavailable. All of these forms of quasi-reality are represented in the growth of consciousness; and to give the so-called ontological form characteristic of knowledge a monopoly of reality is entirely without justification.

To illustrate further, why has not the ejective consciousness, which reads a mental life into the object of presentation, or the inter-subjective, which considers reality as fundamentally social and as possibly endowed with an over-individual consciousness—why has not either of these as good a right to enter a claim to exclusive authority?

12. If by the metaphysician's logic we mean a procedure that assumes or postulates a reality ontological in its character, separate from knowledge, and in some way responsible for the world and all that is therein, we can not accept it. It makes the recognition of the trans-subjective the exclusive weapon of the theory of reality. The real is something apart from experience. On the contrary, there are as many weapons as there are contacts of mind with objects, modes of apprehending and enjoying objective things—and subjective things as well—and the true way of understanding reality involves the use of them all, for the full realisation of the nature of things.

So we come again to the plain story of conscious process and the meanings it achieves. The logicist and ontologist are, on the whole, not true to psychology. They isolate and reify and ontologise single and partial psychological motives, and pass some one of them off for the whole.

The difficulty of getting any understanding of reality would seem, however, to be increased if, while denying the adequacy of the logical and ontological, and of every interpretation which proceeds by the carrying out of a single psychological motive only, we at the same time hold that it is in conscious process itself that reality is, in some fashion, to reveal itself. If these several revelations are fragmentary and contradictory, and if all trans-subjective resource is forbidden to us, then do we not land in a sceptical subjectivism or a positivism based on the special sciences?

13. These are the alternatives currently allowed to those who deny the rationalist's and absolutist's theses; but it remains to point out a further alternative, which the entire body of genetic research appears to justify: the alternative seen in the recognition of modes of conscious process in which, instead of finding opposed motives contending for place, we see experience establishing, *of itself*, a synthetic mode of apprehension. To our mind, the course of the history of thought makes it plain that the quest for such a mode of experience presents the only hope of a lessened strife among points of view; for in such a mode of process evidence would be present to show that the entire system of experience is expressive of reality, and that only in the organisation of the whole are the respective rôles of this and that function to be made out. We have seen that such a quest has been made in certain directions. Synthetic experiences such as those of the self, religious faith, mystic illumination, have been pointed out. These have at least rendered testimony to the need of carrying out to their legitimate outcome all the hints that consciousness gives as to its unreduced and undivided epistemological calling.

14. Accordingly, we may claim without hesitation for this way of approach to the problem of interpretation, the following three virtues.

First. It pursues the sober empirical method of observation and analysis, controlled by the actual movements and events of the mental life.

Second. It does explicitly, what all the other theories do more or less tacitly or clandestinely: it finds in experience itself the sole means of approaching the real. Reality, in the last analysis, *is what we mean by reality*. Reality apart from all meaning for experience is an absurdity or a mere word.

Third. It is constructive, in the sense that it does not deny the epistemological value of any of the mental functions, or the force of any of the theories which are based respectively upon one or other of the functions; on the contrary, its aim is to discover the synthetic adjustment of their claims within the larger whole.

PART III. ÆSTHETIC IMMEDIACY

CHAPTER XIII

THE INTRINSIC SYNTHESIS, ÆSTHETIC

§ *1. The first Step: imaginative Semblance*

IT remains to make good the statement, already put forward in a preliminary way,[1] that in the æsthetic contemplation of an object experience achieves the synthetic and full apprehension of reality. Our historical note has shown not only that some such result is pointed to, as the necessary supplement to the outcome of reflection in its course up to the present, but also that this type of solution has been indicated in a sporadic way, notably by mystical writers and by those idealists who have valued the integrity of personality. The results accruing to our knowledge from new analyses of the æsthetic function and its objects, and the equally new and important insights into the reality and nature of affective logic, enable us to pass on rapidly to our conclusion. The following pages will have perforce to refer to the treatment of the æsthetic and of affective logic as given elsewhere, for details on these subjects.[2]

1. The æsthetic experience is so rich in meaning that

[1] See the citations made in the Preface.

[2] As stated in the detailed treatment of these two topics, in *Thought and Things*, vol. iii., the new view as to the æsthetic is associated especially with the name of Lipps, and the theory of affective logic with that of Th. Ribot.

we are able to recognise no less than four suggestions of dualistic meaning whenever it is experienced, each contributing, however, to the immediacy of the whole effect. There is, in the æsthetic object, first, the character of imaginative semblance, which suggests the ordinary dualism between idea and fact; there is, second, the character of idealisation, which suggests the dualism between fact and end; there is, third, the character of self-embodiment or personalisation, suggesting the dualism between the self and the not-self; and finally, fourth, there is the character of singularity, suggesting the dualism between singular and universal. All these shadings of meaning are positively present in the genuine appreciation of any work or art. It remains to show, however, that instead of developing themselves, these strains of dualism lose themselves in the rich synthesis of immediate contemplation. With all its varied suggestions, no state of mind is more fully one and undivided than that of æsthetic enjoyment, when once it is fully entered into.

2. No synthesis of motives in consciousness, of a sort to overcome its dualisms and oppositions, can take place, of course, while the motives to these dualisms and oppositions remain fully in force. The difficulties may be glossed over, as in mystic states, or denied, as in *ex parte* theories; but only by their effective removal can place be made for a vital union and synthesis.

This release from the bondage of urgent and divided motives takes place in the imagination—a release from the restraints of the external and from the impulsions of the internal[1]; from the conflicts of will

[1] See *Thought and Things*, vol. i., chap. vi. Various writers, notably the poet Schiller, have pointed out this rôle of the imagination, particularly in play. Schiller goes further in the same direction, suggesting

and intelligence; from the opposition of means and ends, and of idea and fact. What is imagined is, in a sense, in the mind; it is no longer strictly limited to the conditions of rival control which, in the prosaic actual world, impose their oppositions and contradictions upon us.

This is the case not merely in the realm of irresponsible fancy, in which the contents of imagination are unorganised and chaotic, fugitive and valueless; but also, in a more restricted sense, in the orderly products of the constructive imagination. The child's playful fancy, although brought constantly into contact with the things of external fact and the thoughts of other persons, develops its dramatic renderings of life and things in a world of semblance or imaginative reinstatement. The playful imagination is the earliest field in which the possibility appears of a more or less independently developing system of presentations, having the organisation due to the existing external control, but free from that control itself. It has indirect reference to the external, and in the same fashion, on the subjective side, to the inner world; but taken simply for itself, it is a world of a certain freedom. The impulses of free construction work themselves out and the possibilities of personal freedom begin to realise themselves. This develops, on the one hand, into the experimentation of serious scientific method, and on the other hand, into the responsible constructions of the creative imagination in art.

The important thing to note is that here we have

---

that art aids in the formation of harmonious and virtuous character, by opening up an imaginative field in which crude desire and instinct may be directed into the channels of spiritual morality. *Cf*. Höffding, *History of Modern Philosophy*, ii., pp. 133-4.

a third alternative in the development of motives, a realm differing from that of the true and also from that of the good. The true is found in the realm of objective control, that of facts and relations; the good in that of the inner life, where satisfactions are desired and attained. These furnish the fundamental alternatives which run through life and theory alike. But here in the realm of the imagination there is a detachment, a free play of ideas. It affords a sort of common ground to the two realms of serious thought and strenuous action. Both the true and the good become "semblant"; they are set up for imaginative purposes, subject to many of the liberties of a free and constructive fancy. Play presents the preliminary loosening of the bonds of fact and value alike; imagination takes advantage of this, in the interest of science and art.

3. This removal of the limitations presented by the dualism of controls is, however, merely preliminary. The world of imagination, like those of fact and value, enters into new phases with the ripening of the motives of conscious reflection. The gradual hardening of the dualism of the inner and outer worlds, mind and body, brings a new phase also in this third world of imagination, in which the opposition is broken down in a new freedom of construction. The æsthetic takes its place along with the scientific and the practical, doing for them what, at the lower stage, the playful fancy does for the perceptual and active. The scientific imagination constructs hypotheses in the interests of further control in the domain of truth; the idealising imagination erects ideals and postulates in the interests of the achievement of higher practical and social values; the æsthetic releases both of these from their partial and contrasting ends, the true and the good, and unites them

in an interest whose end is intrinsic to the construction itself. It sweeps beyond the truth set up by knowledge as complete and absolute, and also beyond the good set up by the will as full realisation of the self; and depicts the real, perfect in all its aspects, as present in the semblant object itself. How it is able to do this can appear only from the analysis of the æsthetic interest; and by what right it is considered valid for the apprehension of reality, only from the meaning of the æsthetic object, the work of art.

4. Put broadly the question is this: granted that in the semblant constructions of the æsthetic imagination, the direct controls external and inner do not insist upon their oppositions, and that the contradictions of ordinary serious life and practice no longer appear, what further then do we find in this mode of apprehension? Do we have the right to go beyond this negative result—which leaves us in the realm of direct feeling, of mystic trance, or of the mere eulogy of beauty—and see here a mode of constructive mental process, inclusive of knowledge and will?

§ 2. *Æsthetic Interest synthetic in the sense of Intrinsic or Autotelic*

5. The grounds on which we may with confidence hold that in the imaginative semblance of æsthetic contemplation a real synthesis is found have been already intimated[1]; here it remains to gather them up and show their force.

(1) In the first place, in the æsthetic construction an interest is at work which does not terminate upon a new or unfamiliar object, but upon an object which is already

[1] See *Thought and Things*, vol. iii., chaps. x.–xiii.

the end of both the theoretical and practical interests. The æsthetic is not excited by that which is foreign to the other interests; on the contrary, it is concerned to keep their objects intact. It is only in that which is in some measure true, good, or both together, that we discern the beautiful.

The distinguishing thing, therefore, about the æsthetic interest is its end: it seeks the intrinsic meaning of the object, not a meaning foreign to or beyond the object. In this respect it is in contrast with the other great interests, whose objects are instrumental to further ends. In æsthetic contemplation, such interests are arrested in their normal course; each is asked to be content with its gains. In this fashion, the opposition which the exclusive pursuit of any one of them would produce is avoided. The æsthetic interest is motived by a reading of the object as being both true and good, with a meaning which the semblant imagination is charged to interpret.

6. (2) The imagination interprets the object for what it is and stands for; not for what either ideal alone, theoretical or practical, would make of it. Each of these, as ideal, is exclusive; each demands the negation of the ideal of the other, or at least its subordination, as part of its own fulfilment. Absolute truth must be neutral, disinterested, says the intellectualist; absolute value must be personal, immediate, say the voluntarist and the mystic. Both parties make their claim in the pursuit of their respective ideals, but in contravention of the actual reports of experience. The true object is also valuable, and the enjoyment of something does not exclude its being a fact of the external world. Here it is that the intrinsic æsthetic interest enters in and demands the revision of the data with a view to dis-

covering their full and intrinsic meaning and ideal. The true and the good are not merely realised alternatively or in succession, but together, as factors in the larger ideal of the perfect.

In the freedom offered by the semblant imagination, this interest in intrinsic completeness and perfection comes into its own. There is no hindrance from the external, since the demands of fact and truth are respected. So also with the good, the useful, the practical: these are not infringed. The interests of the self are united and renewed in the imaginative construction; the divisions of theory and practice are healed; the way is open to the reading of the object in the light of a synthetic consummation. This ideal takes the place of the others, not by reason of a capricious or merely sentimental tendency to do so, but by the legitimate reading of all the characters, recognitive and appreciative alike, which the object presents. This becomes the true ideal imposed upon the objective data; it gives them their fullest rendering. They are no longer mutilated in the interest of a remote design. Art is synthetic in the sense, exactly, that it rids other interests of their narrow and exclusive pretentions, and causes them to join hands in a common task. Their outcome becomes its outcome. The art interest alone does not distort its data.

One may ask by what right the æsthetic interest supersedes and reinterprets the results of knowledge and practice? To this we may reply: by the right that any more integral and complex genetic process may supersede and include those that are less so. All development shows this.[1]

[1] It illustrates the general principle upon which the theory of "genetic modes" is based: see the writer's *Development and Evolution*, chap. xix.

§ 3. *The Æsthetic Object synthetic in the sense of A-dualistic*

7. The interest of æsthetic appreciation demands, as we have seen, an object that unites the true and the good. It demands, that is, one that illustrates the external control and also allows the satisfaction which comes from the realisation of the inner principle or self. The former is present in the construction of a neutral external content of knowledge; the latter in the absorption of those properties of the object which give satisfaction, as being in some sense contributed to it by the self that enjoys it.

The question of further analysis is then this: does the æsthetic object satisfy the demand for the intrinsic union of the two control principles, subjective and objective, upon which the more superficial dualism of knowledge and practice depends? We might conceive an interest which could find both truth and good in the object, so erecting a further synthetic ideal, while at the same time maintaining this ideal as objective and separate from the subject, the thinker or the producer. It is in somewhat this sense that the Aristotelian absolute is æsthetic; while present to the contemplation of God, it still remains objective. Within the sphere of the æsthetic, the distinction would seem to be reflected in that between the producer, the artist, to whom the work of art is an embodiment of himself, and the spectator or contemplator, who retains in some degree an objective and critical attitude. We have to ask whether the genuine æsthetic interest stops with the erection and contemplation of a purely objective construction?

8. Many recent analyses point to a decidedly negative answer to this question. Detailed expositions

and discussions have brought out the facts of the "personalising" of the object by the movement of æsthetic interest. The facts of æsthetic sympathy,[1] *Einfühlung* or Empathy, now so fully established in fact and discussed in theory, come to fill in the picture. In æsthetic appreciation, the object is read as possessed of the very mental and moral life of the observer, so far as this can stand for other spectators also, not by proxy or by mere sympathy, but by the very conditions of its construction and by virtue of the interest which motives its apprehension. It is fully æsthetic only in so far as it is fully the vehicle of the life of the thinker who sees it and feels its force.

In this æsthetic personalising, indeed, we see the mind's own successful protest against all those movements by which experiences fall into the two classes, representing minds and bodies. In vague and inarticulate form, this protest is repeatedly made before the æsthetic proper is fully reached. The personification or "animation" of nature in primitive culture, the presence of the motives of ejection and introjection[2] in the social and religious life; the postulates of mystical union and direct communion in religious idealism—all these show the demand for some means of healing the

[1] According to the theory of *Einfühlung* or æsthetic sympathy, in every case of æsthetic enjoyment the work of art is found to be endowed with life, movement, mind; and not only so, but the spectator goes on to identify this life within the object with the general aspects of his own. For detailed explanations and citations of authorities, see *Thought and Things*, vol. iii., chap. xi., § 3; and *cf.* chap. xv, § 6, below.

[2] Avenarius, while showing in his theory of "introjection," the necessary development of the personalising motive, nevertheless holds that its dualistic results are mistaken and must be revised in view of "immanental" analysis. See R. Avenarius, *Der menschliche Weltbegriff*, and the exposition given in the writer's *Mental Development in the Child and the Race*, 3d ed., pp. 322.

schism between subject and object, and of asserting the oneness of experience in a whole which is conscious of its oneness.

9. In the æsthetic, this reassertion of unity is secured. The semblant reconstruction is a personification. The æsthetic content is shot through with the control proper to the self. The object is not merely presented to me for my observation or criticism; but in it I find the inner world mirrored; in it I feel my own cognitive and active powers establishing themselves. It is not a world foreign to my own life; for in this world, presented semblantly to my gaze, I find realised my community with other selves.

The lesson of it is that the diremption of controls has been, all the while, that merely of partial functions, of instrumental adjustments, of convenient interpretations. The whole story is not told until there is an entire experience, reintegrating in its own life and movement the several phases of the real which mediated truths and values disclose.

§ 4. *The Æsthetic Ideal synthetic in the sense of Syntelic*

10. With this understanding of the synthetic character of the æsthetic interest and its object, we see in what sense the æsthetic ideal is also one of synthesis. If the interest is a union of motives otherwise divided and contending, and if these motives are now held to a relative completeness in the personal life of the knower or producer, the ideal must be one proper to such an interest: the ideal of a fully achieved and perfected object. If the interest is *sui generis*, pursuing neither truth nor practice in itself, but their permanent and

effective union in a larger meaning, then its ideal will include theirs, in so far as their ideals can join in promoting its own. It can not take up the standpoint or pursue the ideal of either as such; this would be to forsake the beautiful. Æsthetic appreciation finds its ideal in a whole which is a perfect and immediate unity, not a mere composition of parts.

For this reason, æsthetic enjoyment and the pursuit of its ideal seem, at times, to be temporary, sporadic, and capricious, even when their high quality and synthetic character are admitted. The partial interests involved may remain so compelling, urgent, and attractive that the æsthetic synthesis may not fully establish itself. The man of practical affairs is occupied with utilities, arrangements, adjustments of things; the man of science with hypotheses, discoveries, the relationships of things. They have neither time nor training, in view of these pressing interests, to pursue æsthetic enjoyments, which require calm and deliberation. But the artist knows the better way. The æsthetic habit of mind, present in his case, is as truly real, though not so common, as the theoretical and practical habits of mind. The time may come, and the stage of human culture, when the realisation of æsthetic values will supply the last and final ground of judgment, both individual and social, if not of the truth and utility of this or that objective product of thought, still of its comprehensiveness for the insight and enjoyment of mankind. The domination of collective and mystical interests long prevented the development of the logical and utilitarian; man only gradually freed his faculties and attained the *milieu* and interests of science. May it not be that the æsthetic is waiting in turn the evolution which will synthesise the human interests, now so scattered, in the realm of intrinsic values found in fine art?[1]

It is also, no doubt, because of their synthetic character that the states of mind of æsthetic appreciation and production are so difficult to enter into and maintain. Art escapes almost entirely the ken of many; to many more it seems vague, light, and superficial. They do not understand its language. This is true, however, not only of art, but of all complex synthetic mental processes, such as those of penetrating intellectual activity, of prolonged practical effort, of sustained attention. Yet the æsthetic, at least æsthetic appreciation, is not altogether like these; since the content is in a sense given to it, its composition is already effected. Whether the æsthetic value really takes effect upon the observer is a question either of the latter's intellectual understanding of the subject and its suggestions, or of the quality and relative purity of the art by which the subject is treated. Worst of all are those distractions of a practical or intellectual sort which are due to uneducated or trivial habits of mind, or to mistaken efforts to make art the servant of instruction or edification.

But the ideal remains in spite of these difficulties of realising its force in the concrete case. It is an ideal of the completeness and unity of motives of all sorts in a whole which the inner life can absorb and call its own. In and through all the variations and changes which the object may undergo, the observer feels the movement of his own inner imitation and sympathy. The work

[1] In French culture, the progress of æsthetic standards and interests toward this high position in life and thought is to be observed, in my opinion, more than elsewhere. *Cf.* the paper "French and American Ideals," in the *Sociological Review*, April, 1913, pp. 106 ff (printed also in *Neale's Monthly*, April, 1913).

of art is objective to his private thought; but its ideal is the embodiment of the human life with which his own life is identified.[1]

11. This is reflected in certain characters which mark the æsthetic as such, within the whole of the imaginative and semblant functions, and distinguish art from its counterfeit, play.

We have seen elsewhere in what respects art differs from play.[2] In art, the motives of the serious life are not reinstituted fragmentarily and capriciously, for mere recreation or amusement, as they are in play; but systematically and truthfully, in a system in which the judgments of value, appreciation, ideality, are semblantly reconstructed. Art thus becomes in its own sense serious. It is not a mere imitation of the actual; nor is it a caricature of it. It is a re-reading of the actual in the more systematic, perfect, and satisfying form which the abrogation of partial controls and the removal of their oppositions renders possible. The reality of the external is not lost; since the reconstruction preserves the gains of judgment and insight, both theoretical and practical. Nor is reality in the inner world lost or impaired; since the work of art is charged with its very spirit and life. But with this union of essential realities goes the fact of their union in the larger artistic reality of which they are factors. Their separate ideals lapse when their meanings are taken up in the larger end of unity and completeness. As such the ideal of art is "syntelic."

12. This makes plain the reason for another difference between the mere fancy of play and the imagination of fine art—the difference found in the ideality of

[1] See below, chap. xv., sects. 28 ff.

[2] See *Thought and Things*, vol. iii., chap. x., § 4.

the latter. Fine art always idealises. It reads the given in terms of its spirit, and carries it on to completion. In this the character of the imagination reveals itself in its prospecting and interpreting aspect. Not only does the æsthetic interest have an ideal, but its ideal is that which the imagination is able to make out of the actual content. The æsthetic object not only possesses the quality of completeness or perfection, as far as its materials go, but it suggests the ideal in which all perfections unite and all virtues inhere. Completeness of spatial form, of harmonious colouring, of rhythmic movement, of dramatic arrangement, of truthful relation, of moral quality, even of æsthetic proportion and relation, all merge in that ideal of ideals, the perfect work of art.[1] The character of beauty realised in some special way in each beautiful thing is idealised; completeness goes on to become ideal completeness. We are in the presence here, therefore, of a mode of experience in which the union of the actual and the ideal is reached. Actuality is realised in the completed object, which is a synthesis of the true with the good, and of the self with objective things. In it all meanings of actuality merge and fuse in one. But ideality is also realised in its turn, in the full or ideal completeness of which the essential æsthetic quality of the given work of art is a sample and model. This final sort of synthesis, that of actual with ideal, is brought out more fully below.

§5. *The Æsthetic Reality inclusive and privative*

13. The truth of the foregoing determination of the

[1] Such would be the entire world of contemplation as it would appear to an infinite experience. See below, chap. xvi., § 1.

nature of æsthetic reality is fully confirmed from the side of negation. Æsthetic interest is of the exclusive type; it is "privative," in the general sense of seeing nothing but its own select and appropriate object.[1] Negation by privation is in this respect in sharp contrast with logical negation or denial, on the one hand, and with exclusion by rejection, on the other hand. It is a mode of acceptance so selective that the entire world of possible things becomes non-existent for the purposes of the privative interest.

From this it follows that, from the point of view of the ideal, the positive æsthetic achievement involves, in so far, not only the negation of all other ideals as such, but also the refusal to recognise them as rival or even possible constructions. The completed imaginative whole in which the æsthetic interest would be satisfied is at once fully comprehensive and radically intolerant. This is a result of the inclusiveness of the art-content itself. The entire system of truths, with the sort of reality it contains, and also the entire system of worths with their inherent real values, are both incorporated in their integrity in the æsthetic reinstatment. Only that reference and control are lacking by which, when taken alone, they are partial and mutually exclusive. The reality attaching to the artistic realisation becomes, when once achieved, the sole title to acceptance of any content whatsoever. The very negations of logical relationship, and the oppositions and conflicts of practice, themselves enter into the whole of the positively æsthetic real. In art the discordant notes of lower and more partial processes are blended in a larger harmony.

[1] See *Thought and Things*, vol. iii., chap. x., § 5, and *cf.* above, chap. vii., §§ 4, 5.

14. While, however, this abrogation may be real in respect to the modes of negation attaching to partial processes such as those of knowledge and practice, the question remains as to the phase of æsthetic negation by which the positively unæsthetic is recognised, the æsthetically bad, the ugly. We can see how the motive of logical denial, by which an opposition between two equally positive contents is established, may contribute a relative part to the æsthetic reinstatement of a larger system; the negative is relative to other aspects of the system, all of which are taken up in the grasp of the æsthetic interest. We can see also how the practically bad, that which is personally, socially, or morally rejected by a reaction of repulsion or avoidance, may have place also as part of the larger æsthetic synthesis. Art may depict both error and evil. But what shall we say of that mode of the unfit which art itself determines as bad over against itself, the ugly, the hideous?

15. In the earlier discussion,[1] we have seen that, in determining the ugly or æsthetically bad, the æsthetic interest encounters aspects of reality—incoherence, error, bad value of various sorts—established by other modes of process which limit or resist its procedure. The ugly consists of those elements, both objective and subjective, of actual existence which do not allow of idealisation, or which do not allow of the interpretation, as part of personal experience, which æsthetic experience involves. This is to say that the beautiful is established in the midst of a world of things considered as actualities of fact and value, and deals with what it finds.

But for the æsthetic itself, this result is not final.

[1] See *Thought and Things*, vol. iii., chap. x., § 6. In this and the following section (16), the fuller statements of chap. xv., § 5 are anticipated.

It is part of the meaning of the æsthetic ideal to negate, by its intent of privation, any other ideal than itself. It is an ideal of acceptance which does not recognise anything to avoid. When taken in its completeness, therefore, the æsthetic would no longer even recognise the ugly, except as it would exist in the world of fact in which all actualities have a certain incomplete and non-æsthetic existence.

We find something of the same sort, though springing from a different motive, in the moral sphere.[1] Moral exclusion in the world of fact may establish the existence of the morally bad; but the ideal of the good does not carry in its postulation the recognition of the ideally bad, save in the contrasting terms of an ideal of what is to be made non-existent by the assertion of the good. In the religious sphere, also, the ideal excludes the existence of the devil, considered as a bad or malignant deity, although subordinate bad personages are recognised in the sphere of fact.[2]

The difference between these cases, the morally and religiously bad, on the one hand, and the ugly, on the other hand, is this: the æsthetic ideal ignores the ugly, except in the realm of fact, by bringing the beautiful in its own way into a new realm of reality from the point of view of which the ugly, the æsthetically bad, does not exist. This actualising of the positive alone, without the negative, the moral and religious postulates do not succeed in accomplishing. It is the prerogative solely of art.

In its nature as both idealising and realising its ideal in this way, art recognises the æsthetically bad only to deny it. It recognises its existence as fact, by right

[1] *Thought and Things*, vol. iii., chap. ix., § 4.

[2] Above, chap. vii., § 6.

of the inclusion of the subordinate motives of truth and practice; but it denies its right to a place in the realm which art itself establishes.

16. One may ask whether this recognition of the ugly as fact does not vitiate the claim of the ideal to establish itself as the exclusive and final meaning of reality. This point is taken up again below;[1] it is the important question of the rival claims of the idealising and actualising interests over again. Here we may note merely that the æsthetic interest itself does not establish the ugly. The ugly is established through the checking or failure of that interest. It remains true, as it is here our intention to show, that the negative aspect of the æsthetic ideal lends confirmation to the claim of comprehensiveness and finality of the art-meaning. The universe, considered as a work of art, must be a complete whole, not only as including all phases of existence in the realm of fact or actuality, but as also, in virtue of its positive nature, denying to these any further title to reality *save that which it itself lends to them.*

It is of some interest to note that by the active operation of this motive of privation, by which one sees and enjoys just the one thing taken up for contemplation, one may fall into certain mystic states of absorption and trance. These states have often an æsthetic colouring by reason of their tranquillity, their character as being self-contained and sufficient. But they represent artificial growths upon the state of mind found in true æsthetic contemplation. For in the latter the attention is fixed upon the semblant objective content, by which it is informed but not dazed. The self takes on the form of this intelligible content, and

[1] Chap. xiv. § 5.

realises its full meaning reasonably and with entire self-possession.

§ 6. *Æsthetic Reality a synthesis of Universal and Singular*

17. The sharp contrast existing between theoretical and affective logic—between intellect and feeling in their respective renderings of reality—appears in the sphere of what has been called "community," with the contrasted meaning of "privacy." Knowledge is always common property; its reports are for intelligence generally; its objects always have the property of recurrence in objective fact, or in convertible and revivable experience, to which the meanings of generality and universality attach, in different senses, in the course of the development of thought. The result is that the experience in which no reference to a common point of objective and stable confirmation is involved remains outside the scope of cognition. The strictly private as such, the singular, fugitive, unrelated state, the purely affective modification of mind, which has no reference to a thing of existence or fact—these remain, if they exist, the unexpressed remainder, the unknown real, haunting the house of rationalism and logicism, and asserting, in the name of immediate and mystical apprehension and faith, that "reality is richer than thought."

This may be summed up by saying that it is the province of knowledge to reveal the universal and general, the "common"; it reports the singular and private only by those indirect means of description, analogy, and inference, which its characteristic processes permit. But its presupposition, its starting-point is generality.

18. The corresponding examination of the logic of feeling[1] has revealed a series of functions in which a certain logic is also operative: an affective logic. But the embarrassment of such a logic is the lack of generality and universality—the opposite to the embarrassment encountered by cognition. Its starting-point is the residue found in the private, the singular, the immediate, which the cognitive functions leave over. These aspects of the real segregate themselves in the inner control, the personal life, of the individual self. The development of interest, volition, sentiment, results in the apprehension of a series of personal and immediate values which, while carried over to things, are nevertheless established by processes mainly affective and active in character.

This cleft extends through all the things of knowledge and thought; we know the thing, but we feel or appreciate its value. It is of the essence of my knowledge of a thing that my thought, my judgment, my intuition of its character should hold for others by the very terms of the logic of knowledge; but it is equally of the essence of my taste, my feeling of its value, which is most real for me, that I can only in certain indirect and very relative terms forecast through it the character of your taste and appreciation.

Feeling, in other words, has the presupposition of singularity and privacy; it reaches commonness, generality, universality only indirectly by means of ejection and emotional abstraction.[2] It is only in those respects in which feeling reads its generalisations, its emotional constants, into the inner life of others, that something analogous to the general validity of cognition is realised.

[1] *Thought and Things*, vol. iii., chaps. vi.–viii.

[2] *Ibid*, vol. iii., chap. vii.

And even then, it is subject to the rudest shocks of disagreement and lack of conformity.

19. Now so long as consciousness is content to live a double life, a sort of "Dr. Jekyll and Mr. Hyde" existence of double personality, it may be now a logical machine turning out generalities, and again a sensitive plant shrinking from the direct touch of the real. But of course our life is not of this double sort; the self is one, and the two realities built up thus in relative opposition are fundamentally but one. Is there any way, we may ask, that known things can be directly felt, and that felt things can be at the same time also known? Is there any mode of apprehension in which the exclusiveness of each of these functions is lost in the inclusiveness of a new and more comprehensive synthesis?

20. This is just the high rôle of the æsthetic function, as we have abundant reason to believe.

The motives to such a synthesis fully reveal themselves in the realm of experience itself.[1] All the characters of art unite in this full result. The relational objective whole necessary to the work of art—the

[1] We have found them objectively and genetically present in the two main stems of primitive art, which bloom respectively in the fine arts of decoration, on the one hand, and graphic and plastic representation, on the other hand. In the one, the representative arts, all that is cognised as common, general, and universal—every possible object of knowledge, in fine—is depicted in semblant and idealised form. In the other, the decorative arts, which have their root in the motive of self-exhibition, the impulse of the self shows itself in personal interest and effort and in the pursuit of individual value. In representative art, the cognitive impulse presses its end on to completion; in decorative art the affective and self-promotive impulses reveal themselves, making the same claim. And in every work of art, these two factors are united in the unity of a single product which is absolutely unique in its force to each observer, while also universal as the vehicle of human expression. See *Thought and Things*, vol. iii., chap. xiii.

thing, situation, event depicted—reinstates, in its full and undiminished meaning, the commonness and universality of truth, in all its varied forms. Nothing escapes the net of art, from relatively simple colour, sound, and other sensuous relations up to the abstract concepts of mathematics and the undefined affective suggestions of morals and religion. All are cast in the crucible, and recast in the forms of beauty, losing nothing of their truth.

But as depicted, being semblant and ideal, not merely actual, they are open to appropriation by the other great motive of apprehension of the real—the motive of appreciation—by which the entire content is made singular, worthful, immediate, an end and not a means. The self enters in, sweeps out the chambers of empty form and generality, and peoples the house with living and real personalities. The real loses its neutrality as respects value, while retaining its relational form; it loses the multiplicity of general and particular in the unity of the mental life; it loses its separate and fixed mode of existence in the ownership of an indwelling spirit. This is no mere figure of speech; our analysis has made plain its motive and meaning. It might be put in the severe technical terms in which many of the expositions of the matter have been expressed.

21. That we have here a new mode of the psychical, a synthetic immediacy, is shown in the fact that its universality and singularity are not, strictly speaking, the same as those of the functions of knowledge and feeling. I declare the work of art before me to be beautiful by a veritable judgment, having the synnomic force of an assertion of relation. But I mean something which is not merely true of the relational construction upon which my conviction of

its positive character depends—that which all men may confirm—but something which is also true in the sense that it is synnomic of others' appreciation, along with my own. In the one act of appreciative judgment, I read your approval of the beauty of the work of art, as well as your recognition of its objective characters. I mean both that it fulfils the objective requirements of beauty—to which I attribute, more or less consciously, its æsthetic character, and which the experimental science of æsthetics charges itself to discover—and also that you will or should join with me in my judgment of appreciation. It has, in short, the universality of the cognitive construction which arises in the logic of knowledge, and also the universality of the affective disposition or attitude in which the universal of the logic of feeling clothes itself. Although, therefore, the rôle of the cognitive in the constitution of the object as universal is not interfered with, the whole is taken over and given the force of a general and syntelic value. It is judged true for the judgment of all, and it is appreciated worthful for the valuation of all. In it, therefore, theoretical and affective logic meet and flow together.

22. The same is true of the singularity of the work of art; it is of both kinds. Despite its generality, the construction is singular in the sense that, as a product of the artist's inspiration and execution, there is but the one; it has the singularity of any isolated and unduplicated object in the world. But this sort of singularity leaves out the aspect of value which makes an object singular to our appreciation. My friend Brown may be the only Mr. Brown I know; but that does not describe him as my friend. It is with a meaning more intimately felt and enjoyed, that I call him "Brown."

In art, this latter sort of singularity, that of affective logic, is present to supplement the other, that of cognitive logic. Although a single objective thing, the work of art has for me a direct immediacy. It is both "one" for all men, and "one" for me alone. When I exclaim to you on its perfections, I mean both to point out its objective marks as representative and common to our joint perception, and also to appeal to your appreciation of its intimate value; expecting to find that you respond as I do to its power to thrill and move the individual observer. While, with you, I have knowledge of one sort of singularity, I can also sympathetically hope that, as respects the other sort, you feel with me. Here again, therefore, cognitive and affective logic meet. Both render undisturbed their singularities, as they do also their universalities, by reason of the semblant and imaginative character of the objective construction from which the dualism of external and inner control has been radically removed.[1]

This is only to say that, *for the full interpretation of the object, neither of the partial logics is adequate.* Shall we give the epistemological task entirely to

[1] One may ask whether this does not constitute a new dualism within the æsthetic itself; a dualism between community as general and singularity as private. But this is not a dualism, since there is no real opposition. It merely shows that if æsthetic contemplation does reveal the real, it must be a reality in which the aspect of common or general truth, open to all individuals, is included in the larger immediacy of a more comprehensive experience—a conclusion to which we are positively lead later on (chap. xiv., § 1).

One may ask, too, whether the generalisation due to recurrence, as of repeated experiences of the same work of art, does not destroy the singularity of the experience. The reply is that while rendering it an affective or "motive" general of the past, so to speak, it still does not exhaust the novelty and essential singularity of each new case which comes to illustrate this general (see *Thought and Things*, vol. iii, chap. vii., §§ 2 ff.).

knowledge, when to do this is to cut off all the springs of valuation and to destroy altogether the creative and revealing functions of art? On the other hand, can we deny to knowledge its rôle, when the very skeleton of all valuation and artistic production is due to the work of cognitive and relational processes? Evidently the only, as it is the adequate, resource is to recognise the mode of apprehension in which the two great motives of thought and appreciation unite in a full and self-sustaining whole.

23. It would seem to be our duty, therefore, in view of this presumption in favour of the æsthetic consciousness—based as it is upon detailed and convincing considerations—to raise finally the enquiry as to the meaning reality takes on when interpreted from the point of view of fine art. The logical or theoretical interpretation issues, as we have seen, in the view that reality is a system of fixed and neutral actualities, the nature of which is determined by positive science and by experimental and logical demonstration. Its reality is of the sort that cognitive process presupposes. It is the reality reached by the logic of cognition. The practical or ideal interpretation, on the other hand, makes of reality a system of values, as postulated by the appreciating and idealising consciousness, and achieved through the pursuit of ends. It is the reality reached by the logic of feeling and practice. What then, we may now ask, is the nature of reality as reached by the logic of æsthetic realisation, the logic of fine art?

CHAPTER XIV

THE ÆSTHETIC INTERPRETATION

1. The full requirements of synthesis appear in the defects of the historical theories of reality. They seem to be met in what we have seen to be the characters of æsthetic immediacy. It represents a synthesis in respect to interest, object, and ideal—all the aspects under which reality as apprehended may be viewed. It remains to bring out the interpretation which may be based upon this type of experience. Our exposition, here again, may be greatly abbreviated in view of the earlier researches to which reference has been made.

§ *1. Æsthetic Realisation as reconciliation of Presupposition and Postulate*

The interpretation to which the resort to æsthetic experience leads is one based upon direct realising; it is therefore a theory of "immediacy," rather than one of the "mediate" type. This follows from the fact that, in this experience, the real-reference is not outward toward something else, but intrinsic, finding its fulfilment in the object itself. The æsthetic object is self-contained, detached, and privative of all other objects. It would seem then to be possible to distinguish the type of immediacy present in the æsthetic.

2. Evidently it is not of the "primitive" type, since

its motives are extremely complex and synthetic. Even in the earliest forms of art, we find the two great factors of imitation and self-exhibition; and the activity of the imagination is exercised upon materials of all kinds which represent simpler and more elementary processes. Hence the interpretation can not be simply mystical or emotional, of the type which results from the acceptance of reality as that with which our sense-experience brings us into direct contact. The theories of reality based upon this sort of immediacy—called "pure experience," "contact with things," "immersion in the stream of life," direct, anoetic, or a-logical "awareness" of things—get no support from the synthetic movement of the æsthetic consciousness.

There remains over the type of immediacy represented in theories based on transcendence or fulfilment, intuition theories in the main; and those based on reconciliation, personality theories in the main.

We have seen that the former of these, the intuition theories, while taking many forms, unite in recognising intuition as the form, though possibly not the genetic outcome, of empirical knowledge; even what is transcendent and a *priori* must embody itself in knowledge, to which it gives universality and necessity. We may say, therefore, that these theories recognise the immediacy of the ideals of knowledge in theoretical intuition, and those of practice in practical intuition. The difficulty, however, with each of them is that it represents one ideal so exclusively that it is by that fact unable to represent others. The ideals of truth and value remain in their fulfilment different and incommensurable. Truth, as given in the pure reason, is a system of rational implications and presuppositions; value, as given in the practical, is a satisfaction and a postulate. The presupposition and the postulate, in short, remain finally unreconciled.

3. Now the interpretation based upon æsthetic immediacy has no such difficulty; for while the æsthetic ideal is one of fulfilment and transcendence, it is not for this reason that it is immediate. It is not an immediacy due to the exhaustion or abbreviation of a genetic process. Theoretical intuition, as we have seen, arises when the end term, the real control, in a process of mediation, drops away and the ideas present in the mediating context of related terms are taken to be absolute and independent realities. The hypothesis, instead of remaining experimental to discovery, a convenience of method, becomes the absolute axiom or the intuitive truth. On the other hand, in the case of the practical norms or intuitions, the reverse process is in evidence. Instrumental or mediating ideas, the means to the achievement of further ends, fall away; and the ends stands forth as absolute unmediated goods postulated by the active life. In each case, there is an abbreviated process, and an assumption of absoluteness and finality in what is known as intuition.

In the æsthetic, this is not the case. The ideal is immediate, not as representing a final outcome and the result of emphasis laid on either means or ends, but as being in its own nature immediate, at whatever stage and with whatever content. It is immediate in the semblance of the child's imagination and the savage's admiration, as it is in the finished intuitions of the artist and the wrapt contemplation of the æsthetic devotee. Its immediacy, therefore, is not that of completed or mutilated function, but of synthetic and reconciling function; it is the essential union of truth with value.

4. In the terms of genetic interpretation, this may

be put in general form. The reality of æsthetic intuition is not the outcome of a partial mode of apprehension of the real, either theoretical or practical; it is a state of essential and synthetic realisation. It is a union of the presupposition and the postulate, not only in their final intuitive forms but in each of their oppositions in the life of thought and practice. The real of the æsthetic ideal is intrinsic to its own process; and this process has its genetic logic, as both knowledge and practice have theirs. The progression of the semblant imagination, uniting at each stage of mental development the interests of knowledge and action, issues in its appropriate object—not merely something presupposed as ground of knowledge or merely postulated as motive of faith, but realised as intrinsic presence. It is both true and good, but not for either of these reasons real; it is real because beautiful, and by this character both true and good. The presupposition and the postulate are both replaced by the acceptance grounded in contemplation.

§ 2. *Æsthetic Realisation as reconciliation of Actuality and Ideality.*

5. Another type of interpretation, already mentioned as based also upon immediacy and claiming to overcome the difficulties of "mediate" theories, insists upon the synthetic nature of personality. While taking many forms, as we have seen, what is essential to the theory is the claim that in the organisation of the self, taken as a whole, the various partial modes of apprehension of reality are reconciled and completed.

This point of view has many merits, but also certain radical defects. If it be the actual self, the objective "me" of cognition, that is intended, the definition of the self only serves to make more radical the opposition between it and other objects, including other selves. For the process of actualising is common to all objects, selves as well as things; things are as actual as the self. If, on the other hand, it be the subject-self, the "I" that is intended, it may be objected that this self can be given no content; for the subject-self is only the logical presupposition of the experience of the actual self, in which, exclusively, it takes on its concrete personal form. To give it any other reality is to land ourselves in some form of transcendence or mysticism.

If we resort to the super-personal, as postulated in religious reality, we have again the conflict between the actual and the ideal.[1] God, if a concrete person, is actual but not ideal; if an ideal, he is not an actual person—that is the religious dilemma. Whichever is taken for real, the other is merely postulated. The super-personal becomes either merely personal or quite impersonal, a social fellow or an "unknown God."

6. In the act of realisation, recognised in the theory based on æsthetic immediacy, the personal is not in opposition to the impersonal, the self to the "other." The very grounds of the opposition, the opposing theoretical and practical motives which produce the distinction, are removed. The semblant reinstatement of the content of knowledge is, by its very construction, charged with the impulsion of a developing mental life, of a self, of *the self*. The mind achieves a new unity in which its genetic diremptions are cured.

In this respect, as in that last mentioned, the result is not attained at the limit only, at the ideal or as a desired outcome, it is a recurring and ever-present fact

[1] See above, chap. vii., § 3.

of experience. Every time that one admires a beautiful object or contemplates a work of art, one experiences the union of these motives of life which the partial interests of knowledge and action had divided. So that the ideal, the intuition, is of a personal life; which is at the same time an objective content, not a mere postulate or analogy. It is present in each case of the entire series of experiences. Its final statement would be inclusive of all the creative achievements of fine art.

7. In the æsthetic, therefore, the roots of the opposition between actual and ideal are torn up, inasmuch as there is nothing left over in the ideal, after the æsthetic actual is realised; the actual is then complete, both in its objective character and in its subjective value. Even the negative, the unæsthetic or ugly, disappears as we have seen, as soon as the positively æsthetic, the beautiful, establishes itself. If nothing remains to realise, there is no further reason to idealise; or—put in the reverse way—when the ideal is realised, the actual becomes ideal.

§ 3. *Æsthetic Realisation as reconciliation of Freedom and Necessity*

8. A further problem, implied in that of the respective claims of actuality and ideality, concerns the domains of necessity and freedom. Genetically considered, it is a problem of the dualism of controls, which runs through the entire mental life; since the necessity found attaching to the objective and actual opposes itself to the freedom found attaching to the inner life, practical and moral. The entire body of positive science, physical and mental alike, summarising our knowledge of the actual, reduces itself to the recognition

of laws to which a certain necessity attaches. Over against this, the self asserts its freedom of choice, its caprice of action, its moral responsibility for the achievement of freely-chosen and desirable ends. The processes which mediate the actual, in short, introduce the rule of law and necessity, imposed *ab extra* by the nature of the control invoked; while the response of immediate consciousness, in all the realm of values and ideals, is the counter assertion of spontaneity, freedom, and the absence of all law save that which embodies a necessity intrinsic to the ideal itself.[1]

9. The historical attempts to solve this problem have resorted, in most cases, to logical argumentation which in the result simply suppressed one of the essential terms. The ontological assumptions of rationalism and voluntarism alike, proceeding by various devices, suppress the necessity attaching to external control. The world of science and law is made subordinate to that of thought and will; objective is derived from moral necessity. Nature, says Hegel, is objective mind; nature, says Schopenhauer, is unconscious will.

On the other hand, to positivists and materialistic naturalists, freedom is an illusion, the mind is epiphenomenal, only external necessity is real.

The more refined attempts to deal with the problem distinguish between the phenomenal and the real, between degrees of the real, between relative and absolute, between instrumental and final. Nature is the instrument, the tool, of spirit, the stepping-stone to freedom. Freedom passes through the discipline of law and chance.

[1] On the genesis and nature of these two sorts of necessity, known respectively as compulsion and obligation, see *Thought and Things*, vol. ii., chap. viii.

10. Here again it would seem to be impossible to avoid or to gloss over the opposition between nature and freedom, so long as we remain at the standpoint from which this contrast perforce arises. The motives at work to produce the idea of an external world or "nature" are in evidence in certain processes of generalisation and abstraction, operating upon contents which have been isolated in the mental life; and the concept of external necessity is thus drawn from these isolated and abstracted classes of data. But in the very act of defining the external, the self defines itself through other data in which selection, interest, and desire embody themselves over against the "other," the external. This domain, that of interest, is the realm of freedom. So long as these two actualities exist, in inevitable contrast, by reason of the motives present in mental development, neither of them can suppress or ignore the other.[1] Even in the logical mode, when reflection has at its disposal the weapons of discursive thought, and the opposition between subjective and objective is transferred to the realm of inner experience itself, the trans-subjective reference remains, involving the domain of external necessity, "nature" stubborn and defiant.

11. But conscious process itself goes on to show us how the "trick is turned." The matter of actual organisation, the content of the objective system of things, need not have the attribute of external necessity, but may be constituted without it in the domain of the inner life. The necessity of whole and parts, of causal dependence, of ground and sequence, of means and ends—in fact the whole related system of data called nature—may remain intact without that sort of external

[1] Cf. *Thought and Things*, vol. iii., Appendix B, sects. 5 ff.

anchorage by reason of which it is to be interpreted as separate from the mind and foreign to it. Although the world is not in its whole being due to representation, still its being may include the property of being represented. This sort of reconstruction once accomplished, as in the semblant and imaginative consciousness, the entire system of the objective may then be owned, possessed, exploited by the spirit of freedom.

This in fact is the case in the æsthetic experience. The inner life takes over the objective system as a whole. It reads its own life of will, feeling, freedom, into the object, making it the vehicle of its expression and the code of its values; finding in it not the enemy, the opposition, the hindrance it seemed before to be, but the appropriate tool of the realisation of its own ideals. Through it desire informs itself, will regulates itself, impulse organises itself; the world is no longer remote but immediate, charged with the values created or posited by feeling and thought. The life processes forget their quarrels and divorces and go forward in the pursuit of the true and the enjoyment of the good, both present in the synthesis of the beautiful—all because the necessity of the one realm and the freedom of the other are made elements in the realisation of a complete experience. The larger self of contemplation finds the double predicates, united and completed, in the one object.

While necessity thus loses its repugnance to freedom, both together becoming the medium of a restrained and informed impulse of will, the latter, in turn, the freedom of will, loses its character of caprice and irresponsibility. The freedom of play and the license of unbridled fancy are not the true freedom; nor are they what the æsthetic interest allows or guarantees. On the contrary, the true freedom is that of the development of the self, to which the content organised by knowledge contributes an essential instrument and supplies a leading interest.

Freedom thus comes to stand for the development of the self in and through the full exercise of its faculties, knowledge being the informing and directing, as will is the dynamic and moving factor. The organic wholeness of the life of experience is re-established in each pulse of æsthetic realisation. It is a relief from the divisions and embarrassments of determinism, on the side of nature, and from the caprices and vagaries of impulse, on the side of mind. In the comprehensive interest of art, the two controls—that of the lawful organisation of contents and that of restrained and directed activity—show themselves at one.

12. In this result, the important suggestion of Kant would seem to have a certain confirmation, though not on the ground given by him: the suggestion stated above[1] to the effect that in the æsthetic we have an intimation of the unity of the realms of nature and freedom. It is not for the formal reason that the judgment of taste affirms *a priori* the harmony of the theoretical ideas of reason with the practical postulates of morals; for the æsthetic immediacy, while one of completion in the sense of realisation of its own ideals, is not one of formal absoluteness. It realises its own meaning in the most humble and unpretentious concrete thing of beauty and taste. Of course, the universal of form may be read into the experience, as its implicit constitutive principle; but what we actually have is an object conveying a more or less tentative and partial bit of knowledge, and suggesting certain values

[1] Chap. xi., § 4, sects. 17 f.

of the nature of utilities or agreeable associations, which finds itself taken up and made part of the inner life of the self who contemplates it. In this reconstitution, the thing of knowledge loses its actual ties of place, time, and relation external to it, and also its character as the vehicle of the actual realisation of suggested utilities and values; and for both of these the self is substituted—a self which absorbs the content of knowledge and finds in it the natural and adequate channel for the expression of its freedom.

Neither is Kant's suggestion that here, in the æsthetic, we experience the easy adjustment of the formal universal to the concrete particular entirely to the point; for it is not a question of a reconciliation of this kind, but rather one of the curing of a dualism between singular and general, both equally concrete and empirical.[1] To Kant the antithesis is another form of that between freedom and nature; for to him the universal as form represents the realm of mind. But such an antithesis is too formal. The æsthetic reconciliation can be considered as formal in the sense that its ideal of completed form includes both that of formal truth and that of formal goodness; but in any given case it is essentially a synthesis of concrete matter, the matter of fact of nature and mind being taken over and charged with the immediate life of the self.

§ 4. *Æsthetic Realisation an Immediacy both of Reconciliation and of Fulfilment*

13. In what we have just said, it has appeared that the immediacy of æsthetic realisation is one that is found everywhere in the progress of a mind capable

[1] Above, chap. xiii., § 6.

of the exercise of imaging. It is a constant function of imagination to reconstitute the materials of knowledge and practice in forms fit to be taken over by the processes of absorption or personalisation which make them æsthetic. It is in this sense that the æsthetic immediacy is one of synthesis and reconciliation; it unites at each stage the partial objects of thought and will in an objective whole of contemplation.

But it is also true that, as a function of its own kind, it has its continuous development toward an ideal fulfilment. The progression of the beautiful proceeds *pari passu* with that of the true and that of the good; there is an æsthetic logic no less than a theoretical and a practical logic. The work of art is judged to be more or less successful, more or less adequate. Like all other constructions of the mind, it may be judged by objective and comparative standards. While, therefore, it always finds the character of completeness present in its object, still this is a completeness which the interest itself co-operates in producing. The idealisation of the content is an essential motive in æsthetic production; but the content, objectively considered, remains only what it is.

It follows that while the work of art is presented as complete in itself and satisfying to the interest of æsthetic contemplation, still the æsthetic interest itself cannot be arrested permanently upon the one work. It goes on to reach further ends of the same kind. Each complete in its way, the series of æsthetic products admit of greater and greater completeness and perfection. The last word is never said. Although content with one work, the artist at once begins another, in which the result is to surpass the first. So the æsthetic connoisseur, while finding the ideal in each work of art, still seeks it in new and more complete embodiments, with a widening experience and a more disciplined taste.

14. There is here, then, a curious situation, one peculiar to the logic of this function. There is a sort of generalisation of completenesses, of perfections. The single work of art is such because it is read as something self-contained and autotelic, in its own way perfect; but the different cases differ in this very quality of perfection. Like other ideals, that of perfection is postulated as the end of the entire series of its own, cases, while, nevertheless, found actually existing in the meaning of each case.

It may be said that the ethical also presents its ideal, that of virtue, as being partially realised in every virtuous act. This is true, but it does not find in each virtuous act the moral completeness and perfection which would make the one case sufficient for the whole, did no other exist. In the æsthetic, such is the character of the single work: the mind sees, in the one work of art, the ideal as if already realised; the synthesis effected is adequate to the motives then in operation. A reconciliation of divergent interests is accomplished. In the case of the ethical, this is not so; the single virtuous act only stirs up the desire, the moral need, for the further act and the better will. The moral ideal hovers over the agent's aspirations, but constantly eludes his grasp. In the æsthetic, on the contrary, the agent grasps the ideal whenever he has a glimpse of beauty. The thing of beauty asserts its complete and unrivalled perfection and, by reason of its detachment and of its privative mode of negation, denies the reality of everything but itself.

15. In æsthetic contemplation, therefore, imme-

diacy has the meaning of fulfilment or transcendence, although, in the motives involved, it is also one of reconciliation. The impulse toward a full and complete life is fulfilled in that the partial factors of that life are reconciled and united. When our life pursues its special interests of knowledge and will, we are farthest removed from the immediate results of contemplation; thought and action are instruments of discovery and adjustment; but when the demand is renewed for the reconciliation of which these partial interests are capable, the ideal re-establishes itself again and again in the act of contemplation. While we may enquire as to the end of this process of reasserted perfections, the mind itself can not ask this question; for that would be to intimate that perfection had not been attained in the single object of beauty.

§ 5. *Æsthetic Intuition a union of Theoretical and Practical—and more*

16. In the ideal thus found both realised and also renewed, in its own fashion, in the æsthetic progression, the rules or norms of the intuitive æsthetic consciousnesses reveal themselves. They are not mere logical rules, since the interest is not theoretical; nor are they identical with the categorical norms of practice, since right conduct is not the æsthetic end. But the synthesis of these two must be reached, since the æsthetic interest finds its content only in the materials originally constructed under their operation.

It is here, however, that the criteria of art show themselves elusive and intangible. As in the ethical realm the rule of right, terminating in the single experience of good-will, is difficult of formulation, taking on modes that run the risk of formality and casuistry, so here also. Yet the moral imperative requires, by the very conditions of the logic of practice out of which it arises, that common consent, synnomic conformity, shall attach to its decisions. In the case of the æsthetic, this is the more true, since in it the factor of cognitive mediation, of truth and objective relation, is so much more in evidence. The rules of truthful representation are not abrogated in art, nor are those of practical good. One may criticise the truth of the work of art, and also investigate its morality, both quite legitimately; and the artist assists with complaisancy at the criticism and investigation.[1] But there must be something more, some rule, norm, quality over and beyond these, which secures the universal approval of the product as beautiful. What then is it?

17. It can only be one thing, if our analysis is correct; it can be found only in the one aspect in which the work of art is something more than an object of theoretical and practical interest. It is the embodiment of a personal life in the course of its exhibition and realisation of itself. In the work of art each man, or every sensitive man at least, finds himself at home. He is not now concerned with the intellectual interest of finding something true; nor with the practical interest of finding something good; but with the self-realising interest of finding something that absorbs and completes the self. After straying in the fields of exploration, and struggling in the morasses that lie before the palaces of duty, he yields himself to the spell of the home, where everything is familiar, where everything is wholesome, and above all where everything is his own.

[1] On the objective marks and the canons of art, see *Thought and Things*, vol. iii., chap xii.

§ 6. *Æsthetic Reason and Absolute Beauty*

18. In view of what precedes, the theory of "æsthetic reason" comes, in our interpretation, to stand beside that of "theoretical reason" and "practical reason." In the former of these two, we recognise those principles that seem, in Kantian terms, to constitute the very framework and structure, the necessary principles of all knowledge. But they arise, as we have seen, when the mediating content of ideas, in the ordinary process of knowledge, takes on the marks of generality and necessity and assumes the character of absoluteness. The properties of common validity and rational interdependence, acquired in the course of experimental and social experience, are reflected into the individual judgment. Here they pose as marks of universal reason.

Similarly, the norms of social conformity, established in actual life, are reflected into the individual conscience. At the limit of the mediation of the end by the means, the means fall away and the end stands as absolute good, imposing its ideal, the categorical imperative of the practical reason. It is a synnomic rule of conduct, an absolute of value.

19. In the progress of the life of feeling, something analogous takes place. The mediations of thought and action assert themselves with their full force here as always; but besides their independent development toward their respective ideals, they unite in a common development in the immediacy of the æsthetic. The progress toward this ideal is not, like that of the others, a making absolute of one term of a process of mediation; for the two ideals are here fused in a state of immediate realisation. Neither means nor ends fall away; but each makes its own fitting contribution to the meaning

of the real. The content of the æsthetic reason, then—understanding by this term the outcome of the affective life embodied in æsthetic intuition[1]—is not something reflected from the world and society into the judgment and conscience of the individual, but something reflected out of the immediacy of the individual's appreciation into the objective and social worlds. While retaining the synnomic judgments of truth and the synnomic values of duty, it reads its synthetic result, the meaning of beauty, into the world of things and men.[2]

In this reading proper to the æsthetic reason, all organised reality—the true, the good, the human, the divine—takes on the form of a whole of beauty. The æsthetic absolute is born out of the absolutes of theory and practice. Beauty is an ejective, rather than an objective, mark or quality, as all affective interpretations of the objective must be. But it is not, for all that, the mere postulation of an ideal. For that which is postulated already in the good and already confirmed in the true, is now the implicit possession of the mind. The self asserts itself in the whole experience, and the whole confirms itself in the self. The æsthetic reason then is the full reason, the only absolute in the sense in which that term has a consistent and tenable meaning—a sense to be more fully explained in the next chapter.

[1] Reason understood in the sense of a psychical immediacy is really in every case a state of feeling. It is of the nature of an emotional or affective abstract, under which particular theoretical or active processes present themselves on occasion, as described in vol. iii of *Thought and Things*, chap. vii.

[2] It is in this sense that beauty may be described as "objectivised sentiment" (see Gaultier, *The Meaning of Art*, Eng. trans.), that is, sentiment which is informed and morally distinguished, resulting from the union, at whatever stage of culture, of the true and the good, and attributed to the beautiful thing.

PART IV. CONCLUSIONS

CHAPTER XV

PANCALISM: A THEORY OF REALITY

§ *1. What Reality must mean*

1. It is clearly impossible to detach from any given real object the particular coefficient, whether of the external or of the inner, due to the system of processes in which the content has been constructed. The distinction between that which is presented in the mind in some form, and that reference to a sphere or world or class in which it is controlled, anchored, and confirmed —that sphere, in short, in which it has some sort of existence, over and beyond its mere presence as a mental state or image—this distinction confronts us everywhere. The entire history of conscious experience shows the development of the oppositions, the contrasts, the reconciliations of the two great sorts of control or existence, inner and outer, mental and external. The meaning of reality in the first instance is a meaning attaching to an idea, presentation, mental content of some sort, by which its place in one of these spheres of existence or being is indicated.

To say of a thing, therefore, that it is real is no more nor less than to say that it has one or other of these sorts of existence, or that of some combination, modification, or revision of them in view of more refined interpretations; it is not to find some new content or idea or predicate which has a separate existence.[1] Reality is not a thing or content; it is a mode of subsistence in a particular control; it has as many different shades of meaning as there are sorts of subsistence and spheres of control.

2. In our interpretation of the various modes of reality meaning, therefore, the question is not that of isolating one reality as being more real, more solid, more valid than others; for all alike arise in the normal process of experience. It is rather a search for that meaning of reality which brings together the various normal modes of control in the fullest and most comprehensive synthesis. It is not an attempt to explain away any of the sorts of reality, but to give to each its *raison d'être* and proper place in the apprehension of the real as a whole. If consciousness has no such synthetic function, no way of reconstituting its "reals" in a comprehensive meaning, then so much the worse for it: but not for our research; for in that case it is our duty simply to report that such is the case. The negative conclusion on this point and a pluralistic result would be as legitimate an outcome as a positive conclusion and a monistic result.[2]

3. We reach an interpretation which finds in the æsthetic experience, with all its larger connotation as determining the sphere of art and revealing the quality of beauty, such a reconstitution of the various reals in a synthesis of realisation. To this interpretation we have given the name "pancalism." It is a plain and simple proposition, whose grounds have been presented in the two chapters immediately preceding. What we

[1] *Thought and Things*, vol. iii., chap., iv., § 2.

[2] See below chap. xvi., § 3, on the pluralistic alternative.

are justified in taking the real to be is that with which the full and free æsthetic and artistic consciousness finds itself satisfied. *We realise the real in achieving and enjoying the beautiful.*

4. The evident question, however, that comes to mind is this: if all the other modes of reality, control, or existence reached in the development of consciousness are to be taken as relative to one another and to the movement of experience as a whole, why is not the aesthetic mode of realising subject to the same limitation? By what right does it claim more authority than they—than perception, for example, or thought, or idealisation in the sphere of practice? If we describe all the dualisms, oppositions, contrasts, etc., to which the various modes of reaching reality are subject, by the term "relativity," then we may ask of the reality reached by æsthetic contemplation, whether it also is not relative?—and if not, why not? In this modest way, we may approach the question which is the final one in speculative philosophy, as it is of our treatment of genetic logic, the question of the "absolute." In what sense may we call absolute the reality reached in æsthetic experience?

The quest for the absolute in the history of thought, so far as it has pursued a consistent course, has been, when all is said, the effort to determine a form of reality whose being or subsistence does not depend in any way upon a principle, or involve a relation, foreign to itself. The absolute is not necessarily itself without internal structure, movement, or positive character; but it is that which is what it is and remains what it is without any dependence upon or interference from anything else. It is externally, though not internally unrelated; internally, though not externally organised. At any rate, to avoid profitless side-issues,[1] let us take the term absolute in this sense, and enquire how far the æsthetic, rather than any other mode of reality, is in this sense absolute.

§ 2. *Sorts of Relativity*

5. We find attaching to the objects of apprehension three distinguishable meanings of relativity. There is, first, the relativity of the sorts of reality which arise together in consciousness by reason of contrasting or in some way differentiated process: for example, the relativeness of mind and body to each other, at any given stage of mental development. Mind is what it is only in an experience in which body is what it also is then and there; and the reverse. It is the relativity of what we have called in a figure the "cross-section" of consciousness, the contents at a given stage of development being considered as a whole, as an entire organisation proper to that stage. A cross-section of consciousness, at the perceptual stage, shows a certain meaning of body and a certain meaning of mind; the two subsist together in the organisation of contents as a whole. The relativity of "parts in an organised whole" is here in question.

[1] It may be said, in reference to this position, that it is not taken unadvisedly. If one should characterise the absolute in any one of a half-dozen special ways—as the unknowable, the undefinable, the uncaused, the completely rational—one finds in each case that one is taking up a partial point of view, logical, ethical, perceptual, which makes the absolute relative from the point of view of the whole of experience. For example, either will or thought can be considered absolute only by the inclusion in itself of the other, while in concrete experience they are strictly correlative to each other. Carried on to the absolute, to the point at which it absorbs and erases will, thought is no longer thought; and in absorbing thought, under the same supposed conditions, will is no longer will.

6. Second, there is the relativity attaching to an objective content considered as being itself in course of development or as having had an earlier stage and as passing on to a later stage. The progression of the self, that of the external object, that of the logical term, each shows a series of stages established normally in the development of the content looked upon as the same. What I am to-day in my thought of myself is relative to what I was yesterday and to what I am to be to-morrow; the same holds of my thought of you, another personal being. Each of these is not only relative to the "other," at one and the same stage, but relative to itself at some other stage. This we have called, pursuing the same figure further, the "longitudinal view" of consciousness, in contrast with the cross-section view; it supposes an observer looking at the movement of a mind retrospectively and prospectively or in a figure "lengthwise." We have to do here with a "genetic" relativity; as so defined, it is a matter of development.

7. Again, third, we note a further and more subtle sort of relativity arising from the fact of affirmation itself. This has appeared in the discussions of negation in its several modes.[1] The act of acceptance is correlated with that of rejection, and seems to admit, if it does not require, a form of existence as attaching to that which is rejected. So also logical affirmation seems, in certain cases, to lead in the act of denial to the recognition of a sphere of the erroneous; in it is found all that which is negatively related to that which is affirmed. Leaving over the more precise statement for discussion below, we may raise the question of a relativity as between positive and negative mean-

[1] *Thought and Things*, vol. i., chap. ix.; vol. ii., chap. viii.; and above chap. xii., § 5.

ing or, to characterise it briefly, a "relativity of acceptance."

8. Another, the fourth, form of relativity that occurs to us—one that has been a stumbling-block in the pathway of philosophy—is that which attaches to the object considered as being in relation to mind, as being literally an "object" of apprehension or experience, as well as a thing. If reality as such is known, is it not then relative to the self that knows it? This is, in our research, a very central topic, for it is as objects or thoughts primarily, that reality is investigated in genetic logic. The phrase "thought and things" suggests this final relativity, which may properly be called the "epistemological" relativity proper.

Having thus described the senses in which conscious objects generally are relative, let us ask to what degree the æsthetic object escapes one or all of these senses, and so wins the right to be called absolute.

§ 3. *The Æsthetic content a non-relative Whole*

9. As to the first of the senses in which the descriptive term "relative" may be used of a mental object or content—that which signalises the contrast, opposition, or other relation of parts or factors in an organised whole—the situation is clear. In so far as it is true that the æsthetic experience is one of the synthesis of contrasted or opposing terms, just to that extent the relativity attaching to these terms or factors would disappear in the establishment of the whole in which the terms are united. The relativity of self and not-self, that of actual and ideal factors of meaning, that of singular and universal—real as these relativities are with respect to the opposing terms as such—do not attach

to the entire experience in which the terms unite in a construction or meaning inclusive of them all.

10. The question arises still, however, as to the complete isolation and self-sufficiency of the æsthetic object. Is it unrelated as respects other contents and independent of them?

In proceeding to answer this question we have to take the point of view of the inner meaning or intent of the experience, the intrinsic claim of the interest involved. Of course we can point out the actual conditions under which the æsthetic object takes form; the psychical antecedents, the active motives, that condition its appearance. But this is not to interpret the experience itself, its own intent of reality. When we enquire into that, the mode of reality of the æsthetic object, the work of art itself, we find it literally a whole, a detached and isolated construction. While it endures, the partial motives, practical, playful, etc., are for the time in abeyance, awaiting, as it were, the dissolution of the bond which unites all their results in the larger artistic whole. In fact, it has been pointed out by certain investigators that in many cases, notably that of the interest of watching a drama, there is a certain vibration of the mind between the ordinary and prosaic system of actualities and the dramatic situation depicted on the stage. The mind's eye, open in turn to each of the two spheres of actual and semblant, prosaic and ideal, enhances the value of the latter by allowing itself from time to time to lapse into the former. And after the play is over, after the intense concentration of the mind on the depicted situation, there is a violent return, a reaction amounting sometimes to a shock, to the partial interests and concerns of every-day life. To the outside observer, who sees the relations of dependence, cause and effect, etc., this seems to denote a relativity as between the two spheres; but, from the point of view of the æsthetic itself, the accomplished synthesis of art, it is simply the return from the ideal completeness of a fully organised æsthetic whole to the sphere of relativeness, opposition, incompleteness.

That this is the true interpretation is seen in the fact that one can not bring two works of art or two æsthetic situations into any sort of antagonism or comparison *inter se*. If truly artistic, each produces its own special effect; it is appreciated for itself regardless of the other. The two may hold consciousness in turn and they may be variously judged, each for itself; but they cannot be weighed synchronously as two different but related artistic units.[1] The attempt to do this gives the jumble of motives found in certain complex styles—the rococo, the plateresque, certain Gothic motives—which breaks up the æsthetic unity into related parts and sacrifices the value of the whole.

11. So far then as this sort of relativity is concerned—that of the terms or units of meaning to each other, in virtue of which one is said to be relative to the other—the æsthetic object is not relative. It admits no

[1] I may say of two landscapes, for example, that one is more pleasing to me than the other, or is more "to my taste." But this is either to estimate my own feeling or to judge the actual contents of the landscapes. It is not to say that the two æsthetic effects interfere with, limit, or otherwise sustain relations to each other. In passing from one to the other, we feel the need of a readjustment, a reconcentration, by which one of them is completely removed, as a whole, from the mind, before the other is taken up. They cannot become simply related terms in a larger whole, without losing their character as independent works of art. So in pronouncing one portrait "better" than another, we pass upon the portrayal, a matter of accuracy or success of representation. But the poorer portrait may be the superior work of art, both judgments being in themselves unæsthetic.

"other"; it is all-engrossing in its essential interest and self-sustaining in its objective compass. If this is one of the conditions of absoluteness, the æsthetic object is in this respect absolute; it means to be an unrelated and autonomous whole.

§ 4. *The Æsthetic, a non-relative Mode*

12. If what has been said is true of the cross-section of consciousness, it is none the less a fact that, in things æsthetic no less than things truthful and things good, consciousness passes through successive stages of development; and this is not only true to the outside observer, but it is part of the very identity of the state with itself. The interest shifts, changes, rises higher, sinks lower, becomes concentrated or thin. There is a functional process, giving results constantly differing among themselves—in short, a progressive movement, a development of the æsthetic mode of apprehension.

Now, it may be asked, is not each stage or point of this advance relative to the stages before and after it?—and is not this true even for the internal meaning itself? From this "longitudinal" point of view, a relativity of genesis or development seems to be present. Does the æsthetic experience in fact escape this?

13. Reserving for the moment the point of view of the outside observer—it comes up below in the discussion of the relativity of subject and object—we may say of the objective æsthetic meaning or content that for the mind itself, the observer's own mind, there is no such progressive development. Each object æsthetically realised—each genuine work of art—is a unity

which is discontinuous with every other; it has no "before or after" in the movement of consciousness. This is not to say, of course, that the objective content, the situation or event depicted, does not vary; it does vary indefinitely. But in all the variations in the content selected to fulfil the æsthetic interest, the end or proper intent of the interest is always the same.[1] When I say "that is beautiful," or "this is sublime," or "that is in good taste," I express the sort of satisfaction that comes from attaining one and the same fundamental end of interest in different ways; that is, the end of contemplating something intrinsic, something having a full reality, to which the different partial interests of knowledge and practice contribute, each in its own way. As æsthetic meaning, the content of my contemplation has one and the same value, although as a thing or as a good it varies indefinitely.

This is implied in all that has been established as to the intrinsic character of the æsthetic mode of interest.[2] It has no remote end; its end is just to find, realise, fully grasp the whole meaning of the object that is present to it. The interest of knowledge is to extend its object, to enlarge the range of truth; and under this impulse, the interest leaps here and there, prospecting for facts, proposing hypotheses, and seeking information. So too the practical interest pursues utilities, moralities, practical adjustments beyond the situations presented to it and which it uses as means. But unlike these, the æsthetic interest is not continually in pursuit

[1] The interest as function varies in freshness, facility, etc., and its effectiveness varies with training and use, temperament, and information; but in all these conditions, the end pursued is constant.

[2] *Cf. Thought and Things*, vol. iii., chap. x.

of the changing and progressive. On the contrary, it is, in each case, fully satisfied with what it gets. It dies when its object vanishes. It comes again, born *de novo*, when the new work of art excites it. In each case it is a self-sufficient meaning, and in all cases the same meaning: just the added intent which the varying objective constructions need to be found beautiful, besides being found true or good or both.

Such adequacy, lack of relativity, inherent completeness and independence of the "before and after," is postulated in the case of knowledge and practice, but only as an ideal; only at their final culmination in a state of intuition. In intuition, as we have seen, certain terms of the process of mediation—the end in one case, the means in the other—are taken as absolute, in just this sense. They become unconditional, independent, unhypothetical, self-evidencing; matters of immediate and unconditional acceptance. In the case of the æsthetic, this is true not only at the limit, at the ideal consummation, but everywhere and always. It is always intuition; the ideal is always fulfilled. It has the same right to be called absolute, so far as this point is concerned—that it allows of no genetic progression, no before-and-after in its meaning—as rational principles and practical imperatives in the theories in which these are the accepted absolutes.

We should remember, however, to be fair to ourselves, that this is the meaning for consciousness itself, for the point of view of the psychical. The question of genetic relativity, as it appears to the observer who notes the progressive changes both in the æsthetic sense and in the æsthetic ideal, with the progress of knowledge and social organisation, is to be taken account of below.

§ 5. *The Æsthetic, a non-relative Acceptance*

14. Another of the senses in which relativity attaches to objective constructions of consciousness is that, mentioned third above, in which acceptance in whatever form implies a corresponding rejection, affirmation a corresponding denial. In detailed studies of the negative, the various forms in which this relativity appears have been pointed out. Of these, certain general types exhaust the matter for our present purposes.

In the first place, there is the rejection, together with the acceptance, of the active life; the exclusion of that which is unwelcome, undesired, repugnant, over against the welcome given to that which fulfils the opposite conditions. In what sense does this appear in the æsthetic realm? Are there exclusions as well as inclusions, of this positive sort, in the interest which determines the object of beauty or art?

We have seen, in examining the æsthetic negative,[1] that something of this sort appears in the determination of the ugly. The ugly is not merely that which is neglected, or not selected, by the æsthetic interest for its contemplation, but that which is positively avoided as repellent and inhibitive to that interest. While, therefore, as has appeared above, no assertion of a negative or opposed sphere of existence is implied by the æsthetic state itself, still we do find, in the ugly, a determination in connection with the æsthetic interest, in a direction which seems to accept the fact or actuality of the object to which the marks of ugliness are attributed. In the result, then, is not the positively æsthetic, the, beautiful, in a sense relative to the negatively æsthetic,

[1] *Thought and Things*, vol. iii., chap. x., § 5. See also above, chap. xiii., § 5.

the ugly? Does not the æsthetic interest recognise the one as well as the other?

15. Upon full reflection, we found that the answer to this question must be negative: the ugly is not established by the æsthetic interest, nor constructed in a form due to it; on the contrary, it is established in opposition to this interest and in inhibition of it. We have found in the coefficient of the ugly two factors. The ugly is that which resists one or other or both of the motives of the æsthetic interest operative in the semblant imagination, *i.e.*, idealisation and personalisation. The negative æsthetic or ugly would therefore be more properly described as the *positively unæsthetic*, in contrast with the merely non-æsthetic or indifferent.

If this is true, it follows that the form of existence or reality attaching to the ugly is always that of a fact, truth, or worth arising in the operation of some other function and *already determined by it.* The elements, for example, of a waste-heap, which offend the senses of sight and smell, and resist the reading of the whole as an æsthetic object, are apprehended by perceptive and affective processes. They are defined antecedently to the rise of the interest which would find the object beautiful. The sort of reality attaching to the heap is, for æsthetic contemplation, one of a partial and incomplete form of existence. From the point of view of the æsthetic interest and its negative aspects of negation and privation, the object has no existence at all as an æsthetic whole; though it may have, as all other objects have, certain existence values of fact and worth, as long as the æsthetic interest is in abeyance.

16. It is interesting to compare this case with the analogous cases of the logically false and the morally bad. The logically false is that which is denied by the

judgment; it is the failure of an attempted assertion.[1] The morally bad or evil, on the other hand, is that which fails to fulfil, or which positively offends, the ideal set up by the moral will.

In the case of the logically false, the proposition that does not win assent fails thereby to get reality in the logical sphere; fails, that is, to be taken up into the system of truths. Logically then it is unreal. But that does not prevent the elements—perceptual, conceptual, affective, or other—from still being existent and in so far real, in the spheres in which they are separately determined. "Cows are bipeds" is not true; there is no such reality as two-footed cows; but both cows and bipeds still exist in the realms, taken separately, in which these classes have arisen. The conclusion is that here, as in the case of the æsthetic, the realm in which the denial goes forth does not admit, in its mode of reality, certain objects which, in simpler and genetically earlier modes of function, were constituted as real. This means that in the succeeding and more synthetic mode, that of the true or that of the beautiful, as the case may be, something is demanded over and above the qualifications which establish reality in the preceding and simpler modes.

17. Similarly, there are certain cases of the morally bad, which represent that which does not fulfil, or that which positively resists, the ethical ideal: acts and dispositions which can not be taken up and constituted in the sphere of moral reality, but which resist the attribution of moral predicates.[2] This may be true, although

[1] Cf. *Thought and Things*, vol. ii., chap. viii., §§ 4 f. and chap. xiii., § 3.

[2] Such are uninformed, impulsive, unreflective acts, in contrast with acts of deliberately wrong intention, which represent a bad will in a positive sense. These latter constitute the true sphere of moral evil.

we can not say that these bad acts and dispositions are within the ethical sphere. They show simply that some things which have reality in other spheres are not available for the constitution of the morally good. They are real enough from the point of view of fact, physical, mental, and social; they resist and limit the progress of the moral ideal: but in refusing to accept them, or in positively condemning them, the moral sense does not give them recognition in the organisation of the positive reality which the moral ideal postulates. In short, the ethical reality consists in a positive organisation of wills and values. As we have seen it issues from a specific mode of interest and practice. That which is not taken up in this organisation is left in the sphere in which its reality is otherwise established; it is real in the mode proper to it. We may condemn an act or an attitude of will as being inconsistent with our moral ideal; but, in admitting its reality, we may be taking the point of view not of the ethical, but of the factual; not of moral reality, but of mere existence.[1] In these cases, the morally bad is analogous to the æsthetically ugly. Each represents the unfitness of certain materials offered to it for its peculiar mode of construction; but neither asserts the negative reality of these materials within its own system of organisation.

18. But now an essential difference appears, which makes the æsthetic peculiar, and rids it of its last vestige of relativity in this regard. Unlike logical denial and moral rejection, an æsthetic assertion, when

[1] Yet immoral acts as such and immoral persons certainly exist. A bad will is that of a person who knowingly acts in a way to violate his own feeling of the good. Unlike the cases mentioned, these acts and persons do enter into the moral sphere.

once established, isolates its content entirely from entangling alliances, both positive and negative. Its positive assertion renders its content privative, not merely exclusive. It blots out everything except just what it holds in its embrace.

This neither the logical nor the moral can do, for reasons inherent in their nature. The logical denies the attribution of a predicate drawn from one sphere or class to a subject belonging to another, while however recognising both as legitimate within the entire logical world of discourse. Both cows and bipeds are real within the system which the logical interest seeks to extend. It can sometimes be said that these logical classes are irrelevant, or in logical opposition, to one another, never that the assertion of something about one deprives the other of all right to exist. To define cows as quadrupeds does not destroy the bipeds. And the same is true of the ethical. The assertion of the goodness of an act is in so far the assertion of the relative badness, or unfitness at least, of the contrary act; but it never implies the obliteration of the latter from the world of real predication, from the domain of values in which moral predicates find their application.

19. In other terms, logical denial and positive moral rejection are cases of exclusion arising from the establishment of rival or inconsistent classes, one cognitive, the other affective; while æsthetic determination, even that of the ugly, does not involve exclusion. On the contrary, when present at all, it establishes a single and unique class, the very establishment of which effects a synthesis that abolishes all other classes as such. As our detailed discussions have shown, æsthetic assertion alone is purely and entirely privative.[1]

[1] See *Thought and Things* vol. iii., "Interest and Art," chap. x.

This being true we may confidently conclude that the sort of relativity attaching to assertion with negation, and to acceptance with rejection, does not attach to æsthetic appreciation. It is then in this sense also absolute.

§ 6. *The Æsthetic non-relative as respects the Relation of Knower and Known*

20. The last form of relativity pointed out above as attaching to constructions which have the meaning of reality, is that which arises from the relation of the thing known to the person or process that knows it. There is in all knowledge the relation of knower and known; for the knower is in all cases present when any sort of object is apprehended. This relativity, therefore, seems to lie behind all the others.

While this is the case, and while this truth has become the assumption of modern speculative thought, it is still true that the ideal of knowledge—an ideal which imposes itself with ever greater clarity and force with the progress of science—implies an objective system of truths and facts which are what they are apart altogether from knowledge. The assertion of the ontological point of view is in so far common to the metaphysician and the "positive" scientist: reality simply is; it is what it is in its own inherent right and nature; it is not that which is known or that which owes anything of its existence or being to knowledge.

21. This becomes then the great antinomy of knowledge, this reference to something foreign to itself—a reference which, if taken seriously, would cut off the process of knowledge from any participation in the reality of its own objects. Setting up external

things as appearing in his knowledge, the knower nevertheless declares that they are entirely independent of his knowledge.

In our earlier discussion,[1] we adopted the knower's point of view in opposition to that of the metaphysician; the psychical, that is, in opposition to the ontological; we now have to ask the question as to just what this point of view implies in the interpretation of objective reality.

22. What has just been called the antinomy of knowledge, in its objective reference, takes on two forms. In the first place, there is the intent of the object to be independent of the individual process in which it is made up or constructed. The individual knower finds it part of the meaning of the thing that it should be foreign to his perception or knowledge; it has to him a control that is in its meaning foreign and independent of him. This we may call, in so far as it implies the denial of any relation between the thing as such and the merely individual knower, the "extra-psychic reference." This reference, so understood, extends to the perception or apprehension of other persons by the individual; they too are external or foreign to the person who apprehends them.

23. The other case is that in which this implication of foreignness or independence extends itself to all knowers or perceivers alike, so that the thing or object becomes completely isolated from all conscious processes. The extra-psychic reference is taken to hold for all; it is synnomic. An ontological thing or truth exists, physical, spiritual, or whatever its kind, separate from the processes of individual and common apprehension alike. This reference also holds for the re-

[1] Above, chap. xii., § 2.

lation of minds *inter se*. As being thus independent of all psychical process, its objects are said to have the "trans-subjective reference."[1]

We have, therefore, these two distinguishable elements in the entire implication whereby knowledge discovers things apart from itself: the "extra-psychic" and the "trans-subjective."

The question of interpretation then arises, as to which term of the antinomy established by knowledge—the terms of which are the relativity of the object to the individual knower and the independence of the object from the knower—is to be accepted as finally characterising reality. Is the real, in its ultimate form, to be considered as an element in personal experience, or is it to be considered as existing *per se* apart from experience? So stated, the question concerns primarily the validity of the extra-psychic reference; but it is usually assumed that with this goes also the validity of the trans-subjective reference. If the object is really independent of any one observer, it is said, it must be independent of all.

24. We have already seen why it is that knowledge can not solve this antinomy of its own making. In the first place, every attempt it makes to interpret the object as independent of itself, gives a result for which knowledge is still more or less responsible, to which in fact knowledge has contributed something. And the question arises at each stage of cognition as to how knowledge can effectively cut itself off from its own products. But more positively, there is a further reason in view of which knowledge must hold to the relativity of its objects to experience. In becoming reflective or

[1] The use of the terms extra-psychic and trans-subjective, as employed here, is suggested in *Thought and Things*, vol. i., chap. i., sect. 10.

interpretative of its own contents, the process of knowledge re-erects them in the form of ideas, systems of data, truths, thus placing them within the theatre of subjective process. The extra-psychic reference becomes a process of conscious mediation. Whatever result its interpretation may reach, all the oppositions and dualisms by which it recognises the foreignness and independence of its objects are established nevertheless under the presupposition of the subjective processes of judgment and reflection. This "secondary presupposition" of the psychical, as we have elsewhere called it,[1] remains to haunt the house of every ontological theory of reality. In this form, the relativity of subject and object remains to the last for apprehension of the cognitive type.

In the matter of the trans-subjective reference, the antinomy is all the more embarrassing, since the synnomic force of judgment and reflection gives increased stability and isolation to the external object, while itself arising very evidently in certain stages of conscious process. One person's illusion may be corrected by an appeal to the perception of others; even hallucination may be discounted in view of the synnomic results of reflection. In this way one of the two externalising references, the extra-psychic, may be antagonised or neutralised by the other. And the reverse is also true. A judgment citing the common belief of all may be refuted by the private experience of one. So the subjective origin and character of both these references to the external is abundantly shown within experience itself.

25. In the case of realities of value, the same antinomy presents itself, but with greater emphasis on the side of relativity, on account of the extreme ambiguity

[1] *Thought and Things*, vol. ii., chap. xv., § 4.

of objective values and the hesitation of the worth consciousness in giving them extra-psychic meaning. In many cases, it is true, the value is attributed to the object as something inherent in it, or at least independent of the individual's choice and preference—the economic value of the diamond is independent of any one person's appreciation of it—but even in these cases the value seems to retain its over-individual or social dependence upon processes of psychical appreciation and selection. The trans-subjective reference is not in force. And in less objective cases, such as the values of immediate contemplation, direct enjoyment, and ideal satisfaction, the factor of subjective and individual choice is so prominent that no means of conversion of the meaning into objective form is found.

We may safely say, then, that while the reality of value has in some cases the extra-psychic intent, it does not generally have this; and that probably in no case does a worth-meaning have that trans-subjective reference which would place the reality of value apart from individual and social experience of the objects to which the value is assigned. It is, in fact, part of the presupposition of the objective and independent reality of things and truths that they no longer contain elements of selective determination. They are stripped of their value, which retains its reference to experience.

26. Our conclusion then is that realities of both sorts remain relative to conscious process, cognitive and conative, in spite of the reference or implication attaching to them as having foreign and independent existence. The presupposition of conscious process hangs about them; and the facts of inter-subjective intercourse give them relations to the social no less than to individual experience.

27. In view of this conclusion, the question comes up as to the presence of these conditions of relativity in the æsthetic consciousness, when objects of either sort, things or values, are made matter of art production or appreciation.

In answering this question, a broad survey of the characters of the æsthetic consciousness again gives us light. It is quite impossible to hold that, in the æsthetic content, the general distinction between persons and things, mental processes and objective existences, is lost. The situations of art are often such as to sharpen this contrast—the tragedies of fate, necessity, law, and the pathetic conquests or defeats of personal love, pity, passion. In art in general the special individual motives are often in strongest contrast to what is general, necessary, mechanical, or dead. The content is recast in the form of imaginative semblance, as we have seen; but there is nothing in this recasting to destroy the distinction, made by conscious process, between the private self and its object. If we are to hold, as seems just, that the results of knowledge and practice alike are retained in their full force in the entire æsthetic meaning, then the distinction between them, which persists in that entire meaning, can not be denied or minimised. The æsthetic then, we may say, does not abrogate the extra-psychic reference of knowledge. The individual does not lose his sense of personality, as over against the æsthetic content. It is clear that, in any case of admiration of a work of art, one does not actually confuse oneself with it, or the elements of personality portrayed in it with those of other works of art.

It is, therefore, not the individual mental life as such that is charged into the object—and of which so much is made in recent analyses of the beautiful[1]—that is, it is not elements which remain individual and ungeneralised. So far as it is the individual mental life, it is, as our earlier analysis indicated, that body of dispositional interests and habits of the individual which allow of generalisation and ejection, and so stand for other minds as well, or for experience in general.

28. But it may be asked, if this is true does not the antinomy remain in the æsthetic, as in the other modes of consciousness, as between the independence and the dependence of the object as respects the knower? If the æsthetic recognises and asserts the extra-psychic reference, in much the same sense as other functions of apprehension—even though, at the same time, it takes the whole object up into the personal life—is there not here a contradiction in the implications of art?

So far as the extra-psychic reference proper is concerned—that which implies the object's independence of the individual's experience—this would seem to be true. But the antinomy is resolved when we come to consider the trans-subjective reference—that which would separate the object from all experience as such. Here we see the importance of the distinction, heretofore largely overlooked, between these two sorts of objective reference.

We have found that the æsthetic, like the theoretical and the practical in their reflective forms, has a synnomic force. The work of art is appreciated as having the same meaning and value for all competent observers. Now in so far as the individual reads a mental life into the thing of beauty, he speaks not for himself alone, but for the community, for all men; for the æsthetic consciousness in general, not for his private

[1] See above, chap. xiii., § 3.

taste and judgment alone. True it does express his private taste; but that is not its whole meaning. He feels that he is the organ of the larger circle, of the universal taste which works in him. The mental life which his sympathetic feeling reads into the object becomes a representative mental life. Its assertion of worth issues from a typical tribunal of judgment, not from a private bureau of opinion.

While therefore the æsthetic consciousness recognises the separateness of the object from experience, taken in the individual sense of the extra-psychic reference, it does not recognise its separateness from all experience, in the sense of the *trans-subjective reference.* This latter is abrogated by the movement of personalisation or *Einfühlung.* The object is given the motive of a typical mental process, although constituted in the mind of the individual, who proceeds to identify himself with it. The synnomic force of the true and the synnomic value of the good are preserved in the semblant reconstruction itself, giving to its mental life, to its "self," the corresponding meaning of a synnomic conscious function. The object of art does not tolerate any strictly private motives or purposes; it is detached from the individual self, at the same time that it embodies what is common and essential to the life of all.

29. The truth of this position is illustrated by the different ways our personal sympathies are attributed in art. Our sympathies go out to living characters rather than to dead things, to the person rather than to the animal;[1] and when it is a question of rival claims

[1] One recalls the story of the child who, on seeing a picture of the Christian martyrs being devoured by lions in the Roman arena, cried out, "Oh, Mama, there's a poor lion that has no Christian!" The incongruity that makes this amusing is just the child's misplaced sympathy.

and motives, it goes to the higher and more ideal rather than to the lower and less worthy. It is the general and universally human, the ideal self, that is typified in the process of personalisation, the individual self identifying himself with this, and holding others to the same identification.

30. We reach then the important conclusion that, as regards the relativity of knower and known, the æsthetic experience, *taken as a whole*, escapes it. It abrogates the trans-subjective reference which would make of the real something separate from experience altogether; for it interprets the æsthetic object as embodying a movement of psychical determination. It gives the work of art the mission to incorporate the movement of mind in its universal and synnomic form. But, on the other hand, it reasserts the relativity of the individual processes which reinstate, within the æsthetic whole, the concrete situations involving knower and known. The knower knows the known, and knows it as being something apart from his knowledge. This antinomy is reasserted in the æsthetic consciousness *within the organisation of the content; but the entire content, carrying within it this antinomy, is not itself subject to it.*

The work of art is, therefore, not relative to any other mode of reality. The æsthetic experience knows no "other" to its own object, no non-experiential or non-æsthetic reality. The relation of subject and object is part of its content; but it is not itself subject to this distinction. In this sense again its object is absolute. It is a reality allowing no other outside it, although inside it there is all the complex organisation that experience of self and the world involves.

§ 7. *The Knower and his Experience: Pancalism*

31. From these explanations, we may see in what sense the reality given in æsthetic contemplation can be called absolute. If the absolute is the non-relative, then the æsthetic reality is absolute in all the senses discussed. It remains, before pointing out certain philosophical corollaries, in the last chapter, to draw the conclusion which is justified as to the point of view from which reality is most adequately apprehended: our answer to the question of genetic morphology.

It appears clear that, by the fusing and transcending of the various modes of existence, subsistence, presupposition, postulation, etc. represented in the relativities now fully discussed—the inner and the outer, the earlier and the later, the actual and the ideal, the true and the false, the affirmative and the negative, the good and the bad—in transcending all these contrasts, the æsthetic consciousness denies finality to each of them as being in an exclusive sense the point of view for the interpretation of the real. If the system of logical truths, for example, is a partial and incomplete reflection of reality, being relative to the system of teleological values which escape its rendering, then any logical character, such as "identity in difference," consistency, organisation in a whole of implication, can not be accepted as sole criterion of the real. They are criteria certainly of the mode of reality secured by cognitive and logical processes as such; but when the point of view of logical apprehension or thought gives place to that attained in the larger synthesis of the æsthetic, the inadequacy of such criteria to evidence the nature of the whole of reality becomes evident.

32. So too with the mode of reality represented in

purposive meaning, moral and teleological, as embodying the principles of will, intention, ideality. Its results and ideals are relative, being in contrast with those of thought, and subject with it to the larger synthesis found in the æsthetic. Hence will and its conservation —its "identity and continuity"[1]—would seem to be inadequate expressions of the real. Interpreted teleologically, reality is in its very nature relative to reality interpreted theoretically or mechanically. To read into the movement of reality as such a principle exclusively teleological or purposive in character, is to abandon the point of view of synthetic realising for that of the ontological postulation of value.

33. So with each of the partial points of view from which reality may be interpreted: all are relative save that in which relativity and incompleteness are lost in a higher immediacy and completeness. This is, as we have now seen, the point of view of æsthetic experience. It is the knower's point of view, as we have described it, when that point of view attains "absolute" validity; when, that is, it stands for the mode of apprehension that is free from embarrassment due to rival and dualistic motives. In it experience reaches its competent rôle, as reflecting, in a full and absolute presence, that which the several knowers, willers, and feelers of its own private and social history have only partially discerned.

So understood, the knower's point of view is not subjective, for it is no longer individual nor in contrast with the objective. It is not private, for its utterance is that of a synnomic or universal experience. It is neither realist nor idealist, in the traditional meaning of that distinction; since its reality, while in experi-

[1] *Cf.* Urban, *Valuation, etc.*, pp. 401 f.

ence, is not in contrast with anything out of experience: experience is all-comprehensive of reality. It is a point of view *sui generis;* and to describe it fully would be simply to write out again the description and implications of æsthetic experience. We summarise it all when we speak of the art-experience or of the enjoyment of beauty; but who of us could say just what is realised in the enjoyment of a work of art, or just why he enjoys it?

34. We have thus filled in the outline of an interpretation or theory, which finds in æsthetic contemplation the organ of the apprehension of the real in its complete, synthetic, and in certain intelligible senses, absolute form. To this theory we have in advance given the name "pancalism."[1] There are, no doubt, other ways of reaching a result which would in some similar manner utilise the æsthetic category; and the resulting theories might also bear this name. Indeed, in our historical exposition, we have described as "pancalistic" certain views in which this goal has been approached by other paths.

It remains to suggest, in the final chapter, certain philosophical corollaries. They are implications of pancalism; conclusions warranted by the results of genetic logic, although not within its scope in so strict a sense as the foregoing.

[1] *Thought and Things*, vol. iii., Preface and chap. xv., § 7. See also the Preface to this work.

CHAPTER XVI

COROLLARIES

§ *1.* *The Nature of Reality*

1. Having now stated the theory of pancalism, as being the interpretation of reality to which our detailed investigations lead, we may point out briefly certain implications or corollaries having a more general philosophical bearing.[1] The broad question, what then is the nature of reality?—is upon the reader's lips.

The conclusions we have reached allow us to suppose that reality is just all the contents of consciousness so far as organised or capable of organisation in æsthetic or artistic form. The individual consciousness is then the organ of reality. The whole of reality would be the entire experience of a consciousness capable of grasping and contemplating it as an æsthetic whole.

The whole is an organised experience, and this experience has the form of a self. If we ask for further descriptive determinations of reality, we fall at once into one or other of those partial points of view from which we lose the vision of the whole, and reach the apprehension only of some special mode of existence or reality—actual, ideal, good, true, or other.

2. As to these special modes of reality, they are not

[1] The reader may turn to *Thought and Things*, vol. iii., chap. xv. (especially §§ 7, 8), where certain of these implications have already been suggested.

to be considered invalid or unreal; since each contributes, in its own way, to the essential content of the æsthetic organisation. The æsthetic reveals something new, something peculiar; but it accepts and reinstates, in its own way, the realities and even the contrasts of the partial modes. The true remains the true; the good, the good; the reals of ideality and actuality remain, contributing the imperative force of right and the synnomic intent of truth. All bring their contributions to the whole. Each is therefore a valid though, genetically considered, a modal and incomplete aspect of the real. The progress of reality is a real progress, the constant achievement of new modes and phases, each having its characteristic æsthetic synthesis, and the whole presenting an ever-widening and enriched contemplation, as the sciences, the systems of practical utility—economic, political, ethical—and the arts of convention and convenience, work out their distinctive contents. Science retains its entire validity, as the theory and instrument of truth; ethics and the normative disciplines remain entirely free to filfil their task; the æsthetic itself has both its scientific and its normative problems, in which patient investigation and searching analysis are needed. All contribute to the general travail and progress of human effort, and all unite to enrich the æsthetic apprehension of the real. The movement of tradition proceeds, the store of intellectual and moral wealth increases, crowned at every stage by the synthesis of art, which renders it, in the full sense, the embodiment of reality.

3. If we take the point of view of the whole, at any stage, reality becomes absolute in the senses indicated. It is not relative to any other whole; and it synthesizes all the relativities of its own parts. The organisation

of contents persists in it; the truths, goods, values of all kinds surviving in the artistic whole of the beautiful. But this does not mean that no further statement, no later achievement, of reality is possible; for the whole is continually and progressively moving on, with the advance of the motives of its several factors. It is just the quality of the æsthetic ideal to reach finality in every statement of its results; but to say that reality is itself finished, in this intent of finality, is to deny the continued efficacy of the motives themselves upon which this very intent is based—the motives of continued reconciliation of active and theoretical impulses and interests.

4. But the question still persists—does a consciousness of the whole, apprehended as a larger æsthetic experience, exist? The answer is clear, since by the whole we must mean that which synthesizes the various existence-meanings. Certainly the consciousness of it exists so far as it exists, and in the same sense. Both things and values exist by the processes of knowledge and valuation which together, in a single movement of synthesis, constitute the experience of the beautiful. The meaning of existence, like all other meanings in experience, undergoes characteristic modifications. The judgment of existence is not the instrument of æsthetic realisation; rather, this realisation is an immediate awareness of the content and of the self embodied in the content. *An æsthetic whole of reality could not be constituted save in a conscious experience.*

We have no right, as we have seen, to accept the ontological conception of existence or reality; the conception of it as being something cut off entirely from experience. By existence we mean that which is reached as the outcome of knowledge, or as the ideal of goodness, or as the confirmed presumption, assumption, or judgment—all modes of conscious handling of data of experience. What is the physical world but what the whole of our physical knowledge finds it to mean—that which our physical experience, broadly understood, leads us to presuppose and accept? So in other realms of existence. This being true, we may say that the experience which presides over and guarantees the partial factors of existence, unifying them all, establishes itself by their organisation in a whole which makes of the world a work of art.

On this point, the postulates of intellectualism and voluntarism are correct, if they are looked upon as methodological assumptions. But neither is capable of superseding the other. Each properly justifies itself on the ground of empirical process of a certain type. But in order to justify its exclusiveness, in denying equal validity to other types of theory, it deserts experience, going over to an absolute of thought or will which is metempirical. This is to cut off the limb upon which it sits, one limb of the entire tree of experience.

5. Philosophical discussion of these positions would, of course, be too extended for our space; they are stated as corollaries of the point of view which our research has justified. To go farther would be to write too long an appendix to the proper topic of the work. So much is legitimate, however, to the problem of real logic, designated as that of the morphology of the real. Answers may be inferred from our main conclusion to many other philosophical problems—answers justified so far as the main conclusion is justified.[1]

[1] It is in place to emphasise the interpretation given to the movement of experience as a whole, as embodied in the theory of "genetic modes" (see the work *Development and Evolution*, chap. xix). See also the discussion of the rival mechanical and teleological theories in Appendix B. to vol. iii. of *Thought and Things*.

§ 2. *The only Alternative—Pluralism*

6. It is in order now to ask what our investigation teaches us in case the synthesis thus suggested is not finally made out. What then becomes the philosophical alternative?

Clearly, there is but one alternative: the field is left clear to a radical pluralism of a realistic and a-logical type. Even a pluralistic spiritualism or personalism would not be justified, since a synthesis based on personality requires the recognition not only of other persons than the one, but also that of things. Personality has both the extra-psychic and the trans-subjective reference. There is in experience no mode of process, save the æsthetic, in which these two references are divorced, that is, in which what is taken to be foreign to individual experience is not also taken to be foreign to collective or synnomic experience. The only sort of a plurality of personalities which would not carry with it a world of things external to all experience, is that which may be constituted within an æsthetic whole.

7. One may, however, ask: why not simply accept a pluralism of realities—physical, mental, moral, artistic, etc.—without any further determination common to them all? Our reply is twofold.

(1) Because to do so would be to take out of experience its objective results, while ignoring the experience itself in which these results arise and have a common significance. These different or plural principles of reality are apprehended in the course of development of a systematic organised mental life; to read them as separate and unrelated to each other and to this mental life would be to deny even the sort

of kinship among them that their common presence in experience establishes. An original or absolute pluralism would contradict even the presupposition of knowledge, upon which the concrete character of each of these realities directly depends.

8. (2) In accepting a pluralism of a radical sort, which takes realities as it finds them—different, disparate, incoherent among themselves—we must in fairness take them all, denying no one of the modes of the real reached in the normal attitudes of acceptance. Among these will certainly be the reality of the beautiful, of art. But in the acceptance of this, as being what analysis shows it to be, we revise our independent acceptance of all the others. If the world is to be artistic, beautiful, *it can not be incoherent, disorganised, radically pluralistic.*

Any pluralistic theory, therefore, that would have a chance of commending itself to reflection, would perforce be a relative pluralism, one of the sort that allows at least the comprehension of the diversity of so-called realities in a larger unity of some sort, such as the unity of experience, or the unity of law. This being granted, the first step is taken toward the conclusion that *the required unity is that of art.*

§3. *Pancalism, a Constructive Affectivism*

9. It remains, finally, to characterise our result from the historical point of view. We have seen that the interpretations of reality, since the introduction of the subjective point of view into modern philosophy, have vibrated between various rationalisms and various voluntarisms, apart from tendencies of a "positive" character, which have recognised certain limitations of

method and so have denied the possibility of a philosophy of reality. In speculative thought systems of rationalism and voluntarism have contested the field. A third point of view, making appeal to feeling, has persisted, however, more or less desultorily, its presentation growing more and more articulate. Its clear formulation is to-day most urgently needed. We have sketched its historical development above under the heading of affectivism.[1]

The result we have ourselves now reached may be called an affectivism; since it denies the competency of either reason or will to serve as exclusive organ of reality, and asserts that an immediate synthesis of functions is found in æsthetic contemplation, a state which may be described as one of feeling. Let us see, therefore, what relation this theory holds to those earlier sorts of affectivism, more or less constructive, to which our attention has already been given.

10. In the first place, considered as a theory based upon immediacy, it satisfies and utilises the demand for direct affective realisation seen in early emotional and religious mysticism. The æsthetic experience is, to the uninitiated, a sort of ecstasy or trance; it has been likened by many to the hypnotic state. The cessation of the motives alike of urgent action and of intense curiosity tends to bring on the calmness and disinterestedness found in simple mystic contemplation.

This side of the experience also gives room for those more definite contributions made to affective theory by early reflective mysticism. In Plotinus, the divine presence and process is not one manifesting itself by causal or logical processes, but one merely presenting itself as a series of actualities, events, data immediately

[1] Above, chap. xi.

present to contemplation. This is true also when the state is one of immediate artistic presence, supervening upon those of rational and practical mediation.

It is in the same movement, further advanced, but satisfying the same demand, that the theories of intuition and faith came into modern philosophy. They rest upon an immediacy considered as embodying what is final and transcendent, something given to consciousness without discursive process, and revealing the ultimate nature of the real. In these views, the directness and actuality of mysticism are taken over and attributed to pure reason, intuition, and faith. In our view, this demand is satisfied in æsthetic contemplation.

11. Again we find in Aristotle, Kant, and Schelling—together with other writers who are more or less explicit—positions in which the æsthetic, or at least a type of feeling, is made the organ of synthesis in which the mystical sorts of contemplation mentioned began to find reflective justification. In many instances, indeed, the resort to mystical and affective states of mind was motived by interests, if not explicitly, yet really, æsthetic in character.[1] The æsthetic has often become the resource, when speculations based upon the motives of partial mental processes were at a dead-lock or had exhausted themselves.

But as mystical contemplation and its later descendants, the various postulates of intuition and faith, did not have the informing and synthetic elements necessary to theory, so the pancalisms of reflective thought have not gone far, owing to their purely speculative and logical character. The determination of the intellec-

[1] This has been brought out in the study of W. D. Furry, *The Æsthetic Consciousness, etc.* (Baltimore, 1908), to which allusion has already been made. See the Preface, above.

tual factor necessary to bring feeling into its true rôle as an instrument of epistemology, has been lacking. Feeling has been left at the level of impulse or passion, or carried over into the empty form of transcendent reason.

What has been needed is the theory of imagination, considered as a function partaking of the nature of cognition and capable of informing the affective interest, while free also to embody it.[1] And it is not feeling alone, but will also, that is to be brought into the synthesis of intuition. The imagination must also be the instrument of the ideals of the will. This began to be prepared for in the doctrine of the imagination of Aristotle and the Italian mystics, reappeared in the theories of art of the Renaissance, and was developed in the doctrines of the schema and of art of Kant.

12. To this constructive development of affectivism our detailed working out of the nature of the schematic imagination, in its assumptive and semblant rôle,[2] may be found to contribute something. The imagination is the instrument of the prospecting, idealising, evaluating, assuming, activities of the mind. It is operative in the sphere of affective no less than in that of cognitive contents and interests; in knowledge, volition, and feeling, no less than in the artistic as such. The newer analyses of the æsthetic, however, show that in it the imagination finds its synthetic and perfect rôle. Its operation in the realisation of the beautiful is not special and relatively unessential, as is usually supposed. In art, the things of knowledge and will, taken up by the imagination, fuse in the immediacy of the values of

[1] As it has been put above, besides the "freeing of thought," there is the "freeing from thought."

[2] *Thought and Things*, vols. i and ii.

feeling; and the two great currents of affectivism, the mystical and the rational, fall together. Whatever may be thought in the result of this attempt to show this, and whatever right we may have to call the resulting synthesis of contemplation æsthetic in this or that accepted sense, still we are persuaded that the materials are present here for a reconciliation of the opposing claims of rationalism and voluntarism, and for a constructive reading of the essential demands of the mystical and intuitive modes of apprehension.[1]

In this sense Pancalism is a constructive affectivism. It shows the way by which feeling may be informed, not remaining blind, but seeing all things *sub specie pulchritudinis*. In this Latin phrase, now explained in detail, the motto of our larger work, τὸ καλὸν πᾶν, is more fully rendered.

---

[1] Professor Urban terminates his valuable treatise on *Valuation*, already-cited, with the following words: "The implications of valuation . . . lead to the claim of priority. Yet since . . . there is always the further implication of an inner truth, since life and experience show themselves . . . more and more capable of statement as a system of truth, there always remains the assumption of the ultimate intelligibility of every value. A still higher form of experience, in which the two claims are equally satisfied, a form of contemplation which transcends will and thought alike, must ever be the goal . . . of all metaphysics. Such a state . . . would indeed be the Beatific Vision."

GLOSSARY OF TERMS

Together with their Definitions, as Employed in Genetic Logic; Accompanied by Illustrations and Instances.[1]

Absolute: not Relative (q. v.). Sustaining no determining relations with anything else.

Acceptance: the attitude of mind which admits something, either as being without question or as putting an end to question.

Cases: Presumption (q. v.): Presupposition (q. v.): Acknowledgment (q. v.): Affirmation (q. v.).

Acknowledgment: the attitude of acceptance accorded to a logical proposition.

Example: I acknowledge the truth that "Lincoln was a patriot."

Actual: (1), existing as a fact or confirmed truth; hence (2), the object of knowledge or of direct apprehension, as contrasted with the postulates and ideals of feeling and will.

Cases: (1) Actual Commonness: the force of being actually held by different minds in common. (2) Actuality theories: those which restrict reality to the facts or truths established by knowledge.

Affirmation: a logical Acknowledgment (q. v.) put in verbal form as an assertion.

Example: the assertion, "Lincoln was a patriot."

Agenetic: not Genetic (q. v.) : not involving real growth or change.

Illustration: mechanical action, such as the action and reaction of billiard balls.

Aggregate: having the Commonness (q. v.) of being entertained by different minds, apart from their consciousness of agreement: common meaning held in common.

[1] The reader may compare the definitions, due to different writers, given in the author's *Dictionary of Philosophy and Psychology*. Some of the terms given here, however, have been proposed since that work was issued and will not be found there. The *Vocabulaire philosophique* of Lalande, prepared for the French Philosophical Society and appearing in their *Bulletin*, has reached the letter P; it is of high authority. Discussions in which some or all of these terms or their equivalents are used, are: A. Meinong, *Über Annahmen* (2 ed.), and *Gegenstandstheorie und Psychologie;* W. M. Urban, *Valuation, its Nature and Laws;* Furry, *The Æsthetic Experience;* Baldwin, *Thought and Things*. See the several Indexes to these works.

Example: common opinion, individually reached. Synonym: Catholic.

Appropriate: fitted to a certain end or use.
Case: appropriate Commonness (q. v.): that sort of common force which renders a judgment fit for general acceptance. *Cf.* Synnomic.

Assertion: the statement of a logical Acknowledgment (q. v.), either affirmative or negative.
Illustration: in a formal debate, one party makes assertion of the affirmative, the other of the negative, of the question proposed.

Assumption: (1), the attitude of assuming something, that is, of treating it as if real or true; (2), that which is thus assumed.
Cases: Suggestion, Proposal, Hypothesis, Postulation, Semblance: see these terms.

Autonomic: self-controlled, self-directed; subject to no authority or law foreign to itself.
Example: the æsthetic interest, as accepting no external interference or control.

Autotelic: having no end or purpose beyond or outside itself; not Heterotelic (q. v.) or Instrumental (q. v.).
Examples: "play for play sake," "art for art sake."

Coefficient: the mark or character which differentiates one mental object from all others.
Examples: "memory coefficient," "coefficient of externality."

Common: in some way involving entertainment by more than one mind. *Cf.* Commonness, and Community.
Illustration: public opinion is common.

Commonness: the property of being Common (q. v.).
Cases: Community (q. v.) of thoughts, Conformity (q. v.) of wills.

Community: (1) the property attaching to judgments and propositions (a), of being accepted by different persons in common ("community by whom"), and (b), of holding good for all persons in common whether accepted by them or not ("community for whom"); (2), the commonness of the acceptance (a) accorded to and (b) due to judgments and propositions.
Examples: (1), the force or import of a proposition as being (a) Aggregate (q. v.), and (b) Synnomic (q. v.); (2), any agreement in belief, whether (a) Actual (q. v.) or (b) Appropriate (q. v.).

Con-aggregate: pertaining to that which is held in common as being common; that is, pertaining to an Aggregate (q. v.) of Syndoxic (q. v.) meanings or knowledges.
Illustration: your and my common understanding of public opinion is con-aggregate.

Conformity: the commonness of consent or active acceptance (a) actually exercised toward and (b) Appropriate (q.v.) in view of a law, precept, or injunction. *Cf.* (a), Syntelic; (b), Synnomic. Examples: the consent of various wills (a) as actually given and (b) as required by the moral law.

Content: that part of an entire Meaning (q. v. sense 1) which is Recognitive (q.v.).
Example: the diamond ring as a recognised object (content), apart from its value, its use for adornment, etc (Intent, q. v.).

Contrast-meaning: one of the terms of a contrast, which, together with the contrasted term, makes a whole single Meaning (q. v.)
Example: inner (-outer).

Control: the limiting, directing, regulative (as over against the constitutive) factor in the determination of anything.
Illustrations: the determination of a physical object has the external control afforded by sensations of resistance; that of ends, goods, objects of desire, has the inner control of appetite, interest, etc.

Conversion: (1) the process of turning a mental object or state into another which it represents or stands for; (2) the resulting testing and confirmation of the former.
Cases: my memory image of a house is converted into a percept when I revisit the actual house (primary conversion); it is converted into your memory or perception when I accept your report to confirm or revise my own (secondary or social conversion); it is converted into judgmental or other psychical terms when I reflect upon, criticise, or otherwise assimilate it to the body of my experience (tertiary or psychical conversion).

Datum: that which is given without co-operation or process on the part of the mind.
Example: a mental object considered as stripped of all Meaning (q. v., sense 2).

Denial: logical rejection; the assertion of the falsity of a proposition.
Illustration: I deny that " the part may be greater than the whole."

Determination: the formation of something, mental or physical, by all the factors which enter into it.
Examples: the (mental) plan of this book as conceived; or the book as a (physical) result.

Dualism: a distinction between two contrasted classes, interpreted as two sorts of existence or reality. *Cf.* Contrast-meaning.
Examples: the dualisms of "mind and body," "self and not-self," "subjective and objective."

Eject (and **Ejective**): another person's mind apprehended in terms of one's own.

Illustrations: the deity, considered anthropomorphically, has been called the "world-eject"; I attribute my motives ejectively to you.

Elucidation: (1), the social expression, as between speaker and hearer, of the Implications (q. v.) of a proposition; (2), all logical matter or truth considered as implying the relation of speaker and hearer.
Example: the logical development of a theme considered as having the force of an oral discourse.

End: that which is set up in the mind as object of desire or of pursuit through this or that Means (q. v.).
Example: riches pursued by means of trade.

External: outside the individual's conscious process, or taken by him to be so. *Cf.* Extra-psychic and Trans-subjective.
Example: the table I write on. Synonym: Outer.

Extra-psychic: apart from or foreign to the individual's processes of apprehension. *Cf.* Trans-subjective.
Example: the reference of knowledge to its object as existing apart from the knower.

Fugitive: pertaining to disconnected, passing, "free" states of mind.
Instances: images of revery, casual fancies.

Genetic: showing real growth or change.
Examples: genetic Series (q. v.), genetic Progression (q. v.).

Heterotelic: having an end or purpose beyond or outside itself; not Autotelic (q. v.). *Cf.* Means, and Instrumental.
Instance: book-keeping as furthering the ends of business.

Immediate: present to consciousness without Mediation (q. v.).
Cases: see Immediacy.

Immediacy: the condition of being Immediate (q. v.).
Cases: (1), primitive immediacy: that of supposed simple sensation or pure feeling; (2), transcendent immediacy: that in which processes of mediation are fulfilled or completed. *Cf.* Intuition; (3), synthetic immediacy: that in which the motives of different mediations are synthesized and reconciled, as in æsthetic contemplation, considered as uniting the true and the good.

Implication: an additional item or items following logically from that which is already accepted or assumed.
Illustration: the falsity of the negative is an implication of the truth of a universal affirmative proposition.

Individuation: the mental process of finding or treating something as a single thing.
Illustrations: the tree, perceived as a whole, including its leaves, and the single leaf, taken as a separate thing, are in turn individuated.

Inner: pertaining or belonging to the individual's consciousness.
Instances: inner process, inner control, etc.

Instrumental: serving as Means (q. v.) or instrument to something else.
Instance: knowledge used for practical ends (in the theory of Instrumentalism, all knowledge is instrumental in this sense).

Intent: that part of an entire Meaning (q. v.) which is Selective (q. v.) over and above that part which is Recognitive (q. v.).
Illustration: a diamond ring has the intent of personal adornment besides its Content (q. v.) as a recognitive object.

Inter-subjective: subsisting between individual minds.
Instance: social intercourse.

Intuition: (1), the content of immediate apprehension; (2), the content of the higher Immediacy (q. v. case 2) of fulfilled or completed process.

Meaning: (1), any mental object with all its signification; (2), the signification (only) which attaches to a mental object, content, or datum (see these terms). *Cf.* Intent.
Examples, as in the phrases: (sense 1) "that idol is my meaning —what *I* mean," and (sense 2) "the idol means God—what *it* means."

Means: that which serves as instrument, tool, or medium to something else. *Cf.* Mediation.
Examples: money, the means of trade; bribery, the political means; "the End (q. v.) justifies the means."

Mediation: the relation by which one mental content (object, idea, etc.) serves as medium or means to the presence or determination of another. *Cf.* Conversion.
Cases: the memory-image of a house mediates the perception of the actual house (cognitive mediation); the plan to take a walk mediates the excursion (active or practical mediation); the "middle term" of a syllogism mediates the conclusion (logical mediation, also cognitive in character); the intention to commit murder mediates the making of the bomb which in turn mediates the actual murder (voluntary mediation, also practical in character).

Mode: Any phase of existence, considered as being *sui generis.*
Cases: in psychology, the (psychic) modes of self-consciousness, volition, thought; in objective science, the modes of vitality or life, consciousness or mind, community or social intercourse; in genetic theory, the (genetic) modes or stages of a genetic Progression (q. v.).

Negation: rejection embodied in a judgment of denial.
Cases: all negative propositions.

Object (mental): whatever the mind can or does apprehend, attend to, or think about.
Illustrations: in saying "I see the dog" and "I am thinking of

nothing," I make "dog" and "nothing" equally my mental object.

Ontological: pertaining to reality considered as independent of or as owing nothing to our apprehension; that is, as existing *per se.*

Illustration: the "ontological point of view" assumes or posits such a reality.

Outer: see External.

Pancalism: the theory according to which the æsthetic mode of being real, apprehended in the contemplation of the beautiful, is all-comprehensive and absolute.

Illustration: in the contemplation and full enjoyment of an object as beautiful, what we realize includes all the aspects under which the object may be found really existent and of real value; as such, it is unrelated (not Relative, q. v.) to anything outside of itself.

Postulate: that which is postulated. See Postulation.

Postulation: the suggestion of something for acceptance, made without adequate grounds in fact or convincing logical proof.

Illustration: the existence of the moral ideal may be postulated.

Presumption (of reality): (1), the primitive and uncritical attitude of acceptance; (2), that which is so accepted.

Examples: the savage's credulity, the child's acceptance of the reality of persons and things. Synonym: Reality-feeling.

Presupposition: the admitted but unstated ground upon which something rests.

Illustration: the statement "the sun will rise to-morrow" has the presupposition of the continued existence of the solar system.

Practical: having reference to practice, utility, or value.

Examples: the practical interest and the practical life as contrasted with the theoretical.

Pragmatelic: tending to or seeking practical ends.

Examples: the motives of gain, pleasure, utility, in contrast with those of knowledge.

Pragmatic: pertaining to that which is practical in its relatively remote and objective consequences.

Example: knowledge considered as leading to or issuing in practical results. Hence the theory which gives such an account of knowledge is called Pragmatism.

Private: peculiar to or possessed by one individual only.

Examples: dreams, immediate states of feeling.

Privation (and **Privative**): the Rejection (q. v.) of all that is not included in a selective Determination (q. v.).

Cases: privative or exclusive interest: the child's rejection of all dolls but "Biddy"; privative negation: a judgment embodying privation, such as, "nothing is worth while but friendship."

Progression: a genetic movement or process involving a series of stages or terms called genetic modes, which are qualitatively *sui generis*. *Cf.* Mode.
Examples: the progression of self-consciousness, that of feeling.

Project (and **Projective**)**:** a mental object at a stage of apprehension earlier than that at which the dualism of inner and outer is reached.
Illustration: both physical objects and persons, as present to the infant's apprehension.

Proposal: a statement suggested verbally to some one for his acceptance.
Example: any logical content or truth considered as suggested to a hearer by a speaker.

Psychic (-al): belonging to the one person's immediate conscious process. *Cf.* Psychological.
Illustration: the psychic "point of view" is that of one's simple awareness of his own psychical or mental events.

Psychological: belonging to the mind as made object of observation. *Cf.* Psychic.
Cases: (1), the observation or interpretation of one mind by another; (2), reflection upon one's own mental processes as if they were those of another; both represent the "psychological point of view."

Psychonomic: pertaining to influences, forces, and conditions which limit, hem in, interfere with, or otherwise serve to Control (q. v.) the mental.
Instance: brain processes considered in relation to the mind.

Public: having the commonness of being recognised by different minds as current or social. *Cf.* Social meaning, under Social.
Instances. Public opinion: that which is recognised by the individual as the common thought of the group. Public meaning (*e.g.*, the self): that thought (*e.g.*, of the self) which is common to all, and of which all recognise the common or social character.

Reality-feeling: the primitive undisturbed sense of reality. *Cf.* Presumption.

Recognitive: (1), subject to recognition; hence, (2), belonging to the cognitive rather than to the conative or affective functions. *Cf.* Meaning, sense (1).
Examples: of (1), a memory image subject to recognition; of (2), a statement of fact as embodying knowledge.

Rejection: (1), the refusal to accept a thing as satisfying or true; (2), opposition, actively put forth.
Examples: of (1), refusal to call the unæsthetic beautiful (*cf.* Privation), or the false, true (*cf.* Denial); of (2), opposition to the morally bad.

Relative: sustaining relations with something else, and in some degree or manner determined by these relations. *Cf.* Absolute.

Schema: a determination of mental content, by the imagination, as a Proposal (q. v.) or Suggestion (q. v.)

Examples: "make-believe," Semblant (q. v.) objects generally, as set up by the imagination.

Schematize: to set up as Schema (q. v.) of the imagination.

Examples; to suggest something playfully, or as an artistic combination, or as a proposed logical relation.

Selective: (1), due to or determined by processes of preference, interest, choice; hence (2), belonging to the affective and active rather than to the cognitive functions. *Cf.* Meaning.

Illustrations: (1), anything desired is a selective meaning; (2), interest is a selective function.

Semblance: the character by reason of which an object is Semblant (q. v.).

Illustration: play has the semblance of reality.

Semblant: set up consciously in the imagination as if real.

Examples: the "as-if" situations of play and art.

Situation: any more or less complex set of conditions.

Instances. Social situation: a given body of social relationships; Self-thought situation: the social relationships implicated in the individual's thought of self.

Social: pertaining, referring, or belonging to two or more minds which sustain relations to one another.

Instances. Social intercourse: typified by the "speaker and hearer" relation; Social meaning: common knowledge, in the sense of Syndoxic (q. v.), recognised by all as common property; Social situation: the body of relationships subsisting in a group.

Socius: (1), the individual's thought of self as involving others; (2), the single self or person considered as involving others in the conditions of his thought.

Illustration: the impossibility of having strictly private thought reveals the socius.

Suggestion: anything present in mind, having the character of a proposal or proposition presented for acceptance, action, or belief.

Example: the suggestion that America existed made by Columbus to Catherine.

Syndoxic: having the Commonness (q. v.) of being entertained by the thinker as common to himself and others.

Examples: conventions, opinions, customs, recognised by the individual as common to himself and others.

Synnomic: having the Commonness (q. v.) of binding force upon all, in the mind of each. *Cf.* Appropriate.

Examples: truths and duties, held by one person to be equally true and binding for all.

Syntelic: having the commonness attaching to ends or purposes entertained in common by different minds.

Illustration: the common intention of different men to get rich.

Teleological: (1) pertaining to that which is intentional, volitional, end-seeking; hence (2), pertaining to that which develops by some inner impulse or motive, as if it were end-seeking.

Illustrations: (1), teleological process: such as the development of an interest; teleological meaning: such as an end held in view, or a system of such ends; (2), vital processes—in appearance tending to an end.

Theoretical: having reference to knowledge, truth, or science.

Examples: the theoretical interest and life, in contrast with the Practical (q. v.).

Trans-subjective: apart from or foreign to all mental processes of apprehension, both individual and social. Cf. Extra-psychic, and Ontological.

Example: the reference of knowledge to its object, as being isolated from all apprehension.

Trial-and-Error: (1), the procedure called try-try-again; (2), the same controlled for scientific purposes.

Illustrations: (1) the child's procedure in learning to walk; (2), the scientific man's method of eliminating rival hypotheses by experiment.

Will-to-believe: readiness to accept one alternative in preference to others, in the absence of logical proof or strong rational presumption.

Illustration: the preference for theism as against atheism, or the reverse, neither being considered proved.

APPENDIX A.[1]

In order to recall to the reader the place of Genetic Morphology in the general scheme of topics covered by Genetic Logic, the following outline of headings originally proposed, in the first volume of the work *Thought and Things*—all of which will have been treated with the issue of the present volume—may be reproduced here, as follows:

GENETIC LOGIC.

A. *Functional Logic* (as in Vol. i. of *Thought and Things*.)

B. *Experimental Logic* (as in Vol. ii. of the same work.)

C. *Real Logic:* i. Genetic Epistemology (as in Vol. iii. of the same work.)

ii. Genetic Morphology (as in the present text.)

In the division given in Vol. i., Chap. i., § 6 of the work cited, Experimental Logic is made a subordinate heading under Functional Logic, instead of being given an independent place, as is done here.

APPENDIX B.

As intimated above (Chap. xi., sect. 24) Professor Ormond assigns to æsthetic experience an epistemological rôle. (A. T. Ormond, *The Foundations of Knowledge*, pp. 227 ff.[2]) He finds that the category of unity has its origin in æsthetic intuition; and concludes that judgment as such, being a unifying function, is always æsthetic. An æsthetic

[1] *Cf.* above, chap. i., § 4.

[2] See also the allusion to Professor Ormond in the Preface.

value attaches to all the products of judgment, in science and morals no less than in art. In this way, the æsthetic is made a matter of feeling in general, which responds to objective unity. Judgment is a function of feeling, in somewhat the sense of the Kantian *Urtheilskraft*.

While recognising the great interest of this suggestion as to the category of unity, the view that the judgment is always æsthetic seems to overlook two important distinctions.

In the first place, to say that unity as such is æsthetic, even in judgments of fact and utility, is to deny the distinction between such judgments, on the one hand, and those of beauty, on the other; or to make it necessary to seek the criterion of beauty in some further character of the objective content. We can say that all unity is pleasing, without going on to say that it is in all circumstances beautiful. On the contrary, it is beautiful—when it is so—because it embodies an ideal, in which a more detached content and a larger synthesis are suggested. To be beautiful, the property of unity must itself become part of the content; it must be gazed upon objectively. Otherwise, it may remain merely a character of the pleasing object—or of a displeasing one—which is not judged to be beautiful. The waste heap, to cite one of the examples of our text, may have its objective unity, but remain for all that offensive and hideous.

This leads to a second point. The judgment as function does not itself unify a content of experience; it merely acknowledges a unity already suggested or proposed. Judgment affirms or denies the synthesis in which the unification is presented in the form of assumption, hypothesis, or schema of the imagination. It is in the imagination that the unity is engendered, the ideal set up. The judgment, by accepting or rejecting it, produces a certain duality rather than unity—the duality between subject and predicate and also that between the object judged and the self that judges. For judgment always establishes the relation of reflection.

We may say, then, that while the category of unity does represent the æsthetic ideal, still it is only in imaginative semblance, in the schema or sketch of unity present to the self for its contemplation, that this ideal is embodied. The unity of judgment may, however, be so presented. In science and morals, on the contrary, while the unity of content is preserved, the act of judging intervenes to disturb the ideal unity of æsthetic feeling. The unity of objective content as such is merely that of cognitive individuation and recognition, appreciated by feeling in general as pleasing or displeasing, but not always by æsthetic feeling as ideal and beautiful. For æsthetic feeling, the dualism of self and object, established by judgment, must be overcome, in order that the self may absorb its content and live in it.

APPENDIX C

I

The philosophical views of Mr. F. H. Bradley[1] touch closely upon those developed in our work, while at the same time illustrating a different method and reaching different final conclusions. It may be of interest to bring this out in a few sentences.[2]

1. Mr. Bradley holds, as I understand him, the following positions—(1), that absolute reality is present to experience, that is (2), to an experience of immediacy, which immediacy (3), is that of feeling or "sentience"; further, (4), that knowledge, proceeding by judgment, renders the real in forms which are never free from contradiction and relativity, although (5), its ideal is the absolute unity of the Whole. Absolute reality, therefore (6), cannot be identi-

[1] F. H. Bradley, *Appearance and Reality*, 2nd ed., and *Essays on Truth and Reality* (1914).

[2] Naturally it is impossible to do justice, in such a brief statement, to Mr. Bradley's very subtle and original views, or even to be sure of not misrepresenting them.

fied with thought, nor can its method of development or change be that of the dialectic of thought processes (as in Hegel). Similarly, (7), there is no absolute error or falsehood, as there is no absolute truth, attaching to judgment, since all judgment "qualifies" reality in some partial way, and relative truth and reality adhere in all empirical process. Finally, (8), it is in a higher immediacy, hyperlogical in its character, in some sense qualified as personal, that the Absolute is to be found.[1]

2. On all these points our own results support the positions of Mr. Bradley, as different as are the grounds on which the respective conclusions in these several fields evidently rest, and as different as are the interpretations of reality finally arrived at.

3. In reaching his final conclusions, Mr. Bradley takes what has the appearance of a plunge into ontology[2]—purely logical and unexperiential in its motive and result. Absolute reality is a whole, present in the immediacy of sentience, but incapable of further definition or appreciation for what it really is; since the objectifying functions which aspire to render it only succeed in masking it. They misrepresent the whole in asserting and qualifying the parts.

Nevertheless, experience belongs to a "finite centre"—each person has one—before which "the whole universe

[1] In his new book, *Essays on Truth and Reality*, Mr. Bradley gives (pp. 417–18) a résumé from which I quote the following sentences: "We have (1), an immediate felt whole, without any self or object. Next, (2), where we find an object against the self, this opposition is still a content within a totality of feeling. . . . Further (3), the self, although not yet an object, is experienced content, and it is a limited content and is so felt. (4), On the nature of that limited content felt as self . . . there is much of it which from time to time has come before us as objective, and on the other hand there are elements which remain throughout in the background. And all this will be true even of that central group on which our personality seems to rest." In other words, the objective self (and the personality) is a content wherein "that finite centre, within which and before which the whole Universe comes," takes on partial and special form.

[2] *Cf.* above, chap. xii., sect. 10.

comes." The absolute experience, therefore, although one of feeling, is not one of mere uninformed and crude feeling. It is not a sentience without rational and practical content, but a sort of super-personality—by implication at least—which has the logical or metaphysical value of all possible predicates of truth and worth, although not able to disclose these predicates in actual judgments. The whole is present in sentience, but, do the best it can, the whole cannot make itself fully or properly evident in the form of "appearance."[1]

4. From this our own result differs sufficiently, despite much preliminary agreement. We have been able to utilize certain resources which Mr. Bradley has not drawn upon, notably the genetic method, the theories of "community" or social intent of judgment and self-hood, the developments of practical, affective, and æsthetic logic. Instead, therefore, of resorting to the ontological point of view and reaching what is after all a logical postulate—that of a "finite centre" before which absolute reality "comes" in the immediacy of sentience—we go on to a "comparative morphology" of the modes of function which give partial renderings or "appearances" of the real. We find that experience itself discloses its own "more excellent way" of apprehending the "whole." To us, the whole is realised in kind in every work of art; while to Mr. Bradley there is a supposed or postulated ontological whole which is not realised but merely "touched," so to speak, in a sentient contact.[2] It is for this contact—to give it metaphysical meaning—that the "finite centre"[3] is postulated.

[1] I venture so much, although Mr. Bradley's latest explanations on the subject of the "finite centre" are exceedingly difficult. See his *Essays on Truth and Reality*, chap. xiv.

[2] See above, chap. x., sect. 8.

[3] "Within which [*i. e.* the finite centre] and before which *the whole universe comes*" [italics mine]: Bradley, as quoted above. One is reminded of the monads of Leibnitz, each of which "presents" the whole world; but it is easier to attribute a content to a centre which "presents" than to one which merely feels, for the former specifically recognises the validity of knowledge.

5. The difference really proceeds from the radical divergence of method signalised in our discussions.[1] It is instructive to bring this out; for our own method forbids the resort to an ontological point of view, as having superior or final validity. We have found[2] in passing that in case the synthetic reconciliation of thought and practice—of all modes of mediation—both *in and with* a new immediacy, were not to be found in actual experience, æsthetic (as to us) or other, then our alternative conclusion would have to be realistic and pluralistic, rather than rationalistic and monistic (as Mr. Bradley's is). We should have to stop where the experience of reality itself stopped: with the recognition of several modes of existence each equally revealing the real, though partially and under its own limitations. We could not go on to say that there is a whole, withal *implicit*, which the processes of appearance cannot make *explicit*.

6. Mr. Bradley, it seems, when all is said, places experience and empirical method under the yoke of absolutism, by the demand for absolute identity in the real: all difference must disappear in absolute identity; this is the logical and metaphysical ideal. This ideal, however, thought fails to attain, and so failing, suffers shipwreck.

But why this ideal? Why not that which experience suggests, a "unity in variety," an "identity in difference," an organised absolute, not one merely sentient nor one merely logical—one blind, the other empty? Such an organised reality—the very ideal of the æsthetic—is realised in every work of art. It makes all the difference between a full, rich, content of realisation and a thin logical absolute void of content: between immediate realism and absolute rationalism.

7. It is only here also that the justification of monism is to be found; it resides in the demand for unity made by æsthetic contemplation. The intellectualist and voluntar-

[1] See above, chap. xii., sect. 2.

[2] Above, chap. xvi., sect. 2.

ist proofs of monism are unconvincing; each considers its own postulate alone final. It is of beauty alone that the substance of the following sentence, already penned above,[1] can be asserted; nothing analogous can be said of truth or of worth:—"If the world is to be artistic, beautiful, *it cannot be incoherent, disorganised, radically pluralistic.*" Here and here alone, in our view, is the compelling reason for monism, and the certitude, found in immediate experience, of the reality and significance of the "finite centre."

II

That the general conception of Genetic Logic, and with it the term—first employed systematically, so far as I know, in the work *Thought and Things, or Genetic Logic*, and in portions of that work published separately—is gaining currency, is seen in the character of recent discussions of logical problems. The issue is fruitfully joined between the metaphysical, "logicist," and Hegelian conceptions, on the one hand, and the empirical genetic, on the other. In a recent reivison of his work, *Logic*, for example, one of the advocates of the Hegelian form of metaphysical logic reprints his criticisms of the present writer's partial theories of "imitation" and "selective thinking," and states his general reply to "the antagonist," the genetic logician. To him and others of the school there can be no "genetic logic," in a true sense, because logical principles, being absolute and universal, can have no genesis. His criticism takes no account of the systematic developments of my later volumes—including, of course, the present work—and is for that reason ineffective, except as stating his own antagonistic view. In a similar revision of my replies to this writer's original criticisms—soon to be reprinted—I hope to bring out fairly the contrast between these two leading conceptions. In the meantime, the sections devoted to

[1] Above, chap. xvi., sect. 8.

the "metaphysician's logic" in the text above, pp. 223 ff, and in the Introduction to Vol. I of *Thought and Things*, as well as the Appendices to Vols. II and III of that work, replying to other critics, may be referred to, The consequences following upon this difference of method and point of view are even more pronounced than those pointed out just above in the case of Mr. Bradley.

INDEX

See also the "Glossary" of terms immediately following the text, reference to which is not made under the corresponding terms in this Index.

O

P

R

For Product Safety Concerns and Information please contact our EU representative GPSR@taylorandfrancis.com
Taylor & Francis Verlag GmbH, Kaufingerstraße 24, 80331 München, Germany

www.ingramcontent.com/pod-product-compliance
Lightning Source LLC
Chambersburg PA
CBHW060741150926
51501CB00003B/63
* 9 7 8 1 1 3 8 5 2 4 2 3 1 *